Urban Transitions and Dynamics: Cultural, Ecological, Digital and Degrowth

Urban Transitions and Dynamics: Cultural, Ecological, Digital and Degrowth

Guest Editors

Pedro Chamusca
André Carmo
Ricardo Almendra
Patrícia Pereira

MDPI

Basel • Beijing • Wuhan • Barcelona • Belgrade • Novi Sad • Cluj • Manchester

Guest Editors
Pedro Chamusca
Faculty of Arts and Humanities
University of Porto
Porto
Portugal

André Carmo
School of Sciences and Technology
University of Évora
Évora
Portugal

Ricardo Almendra
Centre of Studies in Geography and Spatial Planning
University of Coimbra
Coimbra
Portugal

Patrícia Pereira
CICS.NOVA, ESECS
Polytechnic of Leiria
Leiria
Portugal

Editorial Office
MDPI AG
Grosspeteranlage 5
4052 Basel, Switzerland

This is a reprint of the Special Issue, published open access by the journal *Land* (ISSN 2073-445X), freely accessible at: https://www.mdpi.com/journal/land/special_issues/8OAQBP3PCO.

For citation purposes, cite each article independently as indicated on the article page online and as indicated below:

Lastname, A.A.; Lastname, B.B. Article Title. *Journal Name* **Year**, *Volume Number*, Page Range.

ISBN 978-3-7258-5397-7 (Hbk)
ISBN 978-3-7258-5398-4 (PDF)
https://doi.org/10.3390/books978-3-7258-5398-4

Contents

About the Editors vii

Pedro Chamusca, André Carmo, Ricardo Almendra and Patrícia Pereira
Reimagining Urban Futures: Perspectives on Cultural, Environmental and Digital Transitions
Reprinted from: *Land* **2025**, *14*, 1725, https://doi.org/10.3390/land14091725 1

Chenchen Shi and Jinjing Lu
Unlocking Economic Resilience: A New Methodological Approach and Empirical Examination under Digital Transformation
Reprinted from: *Land* **2024**, *13*, 621, https://doi.org/10.3390/land13050621 5

Le Li and Tao Song
Enabling In-Situ Urbanization through Digitalization
Reprinted from: *Land* **2023**, *12*, 1738, https://doi.org/10.3390/land12091738 25

Qiangyi Li, Jiexiao Ge, Mingyu Huang, Xiaoyu Wu and Houbao Fan
Uncovering the Triple Synergy of New-Type Urbanization, Greening and Digitalization in China
Reprinted from: *Land* **2024**, *13*, 1017, https://doi.org/10.3390/land13071017 44

Yuanhe Zhang and Chaobo Zhou
Effect of Artificial Intelligence on Chinese Urban Green Total Factor Productivity
Reprinted from: *Land* **2025**, *14*, 660, https://doi.org/10.3390/land14030660 68

Xiaozheng Zheng, Shuo Yang and Jianjun Huai
Integrating Ecological Footprint into Regional Ecological Well-Being Evaluation: A Case Study of the Guanzhong Plain Urban Agglomeration, China
Reprinted from: *Land* **2025**, *14*, 688, https://doi.org/10.3390/land14040688 88

Ming Tian, Zhuo He, Jinpeng Wei and Yicong Tian
Empirical Analysis on the Mechanism of Industrial Park Driving Urban Expansion: A Case Study of Xining City
Reprinted from: *Land* **2024**, *13*, 1577, https://doi.org/10.3390/land13101577 110

Zihan Hu, Min Wu, Dan Yang, Tao Luo and Yihao Tian
How Resource-Exhausted Cities Get Out of the Innovation Bottom? Evidence from China
Reprinted from: *Land* **2024**, *13*, 1153, https://doi.org/10.3390/land13081153 124

Fangting Chi and Haoying Han
The Impact of High-Speed Rail on Economic Development: A County-Level Analysis
Reprinted from: *Land* **2023**, *12*, 874, https://doi.org/10.3390/land12040874 146

José Fernandes, Pedro Chamusca, Rubén Lois, Helena Madureira, Juliano Mattos and Jorge Pinto
Tourism-Led Change of the City Centre
Reprinted from: *Land* **2024**, *13*, 100, https://doi.org/10.3390/land13010100 168

About the Editors

Pedro Chamusca

Pedro Chamusca is an Assistant Professor at the Department of Geography of The Faculty of Arts and Humanities of the University of Porto. With a PhD in Geography from the Faculty of Arts and Humanities of the University of Porto, with a thesis entitled "Governance and Urban Regeneration: between theory and practices" (2012), Pedro was the President of the Portuguese Association of Geographers (2022–2024) and is also professionally qualified in Geographic Information Systems (GIS). His professional background includes private consultancy work in GIS, planning, and territorial planning (where he founded a company), and a solid career in scientific research: as a fellow in a European project (Chronotope: Time-Space Planning for Resilient Cities?); as a postdoctoral researcher, he coordinated research on the theme of territorial cohesion, especially in its association with territorial innovation processes (carried out at the Department of Social, Political, and Territorial Sciences of the University of Aveiro); and as a contracted researcher at CECS, where he coordinates (nationally) the inPUT project (Engaging Places and Communities for Inclusive Peri-Urban Transitions), developed in consortium with the Universities of Delft, Antwerp, Vienna, and Porto. He is the author of about a hundred scientific publications, Assistant Editor at the *Journal of Urban Development* and a reviewer in several indexed international journals. He conducts research on topics associated with Urban Geography, Territorial Cohesion, Geographic Information Systems, Governance, Planning, Tourism, and Territorial Planning. He was responsible for national and international initiatives such as the Geography Olympiads (2018–2024) and coordinates/participates in various projects for the application of knowledge in areas associated with strategic planning, territorial planning, territorial cooperation, geographic applications of knowledge and decision support, development, and commercial urbanism.

André Carmo

André Carmo holds a PhD in Geography from the University of Lisbon and is currently an Assistant Professor at the School of Science and Technology, University of Évora. He coordinates the Évora hub of CICS.NOVA – Interdisciplinary Centre for Social Sciences, where he conducts research within the Cities, Environment and Regional Development Group (CICS.NOVA.UÉvora). Additionally, he is a collaborating researcher in the Human Rights, Citizenship and Globalisation group at the Centre for Global Studies of Universidade Aberta (CEG-UAb). His research interests focus on the fields of social and political geography, geographical thought, and urban studies. He has participated in numerous research projects at both national and international levels. Most recently, he co-coordinated the project Habitar Siza: The experience and interaction of residents with the architecture and micro-technologies of Álvaro Siza. He is currently leading the development of the Eco.Inc-LAB at the Parish Council of Encosta do Sol, following his previous responsibility for designing the participatory model for Évora's application for the title of European Capital of Culture. He is a member of the board of the Portuguese Association of Geographers and was recently elected President of the General Assembly of the Portuguese Association of Political Economy. He recently coordinated the volumes Espaço, Lugar e Território: Figuras do Pensamento Português Contemporâneo (Afrontamento, 2022), Sindicalismo no Ensino Superior e na Investigação. João Cunha Serra – uma vida em luta (Outro Modo, 2023), and (Re)visitar a Regionalização 25 Anos Depois (Almedina, 2023).

Ricardo Almendra

Ricardo Almendra, PhD in Human Geography, is an Invited Assistant Professor at the Department of Geography and Tourism at the University of Coimbra and a researcher at the Centre for Studies on Geography and Spatial Planning. He has participated in several European (e.g., EURO-HEALTHY, HEALiQS4Cities) and Portuguese (e.g., GeoHealthS, Smaile, ClimaHabs) scientific projects, currently coordinating the FCT-funded BLINC project and co-coordinating Work Package 4 – Good Health and Well-being – of the EC2U Alliance of European Universities. He is a member of the Steering Committee of the International Geographical Union's Commission on Health and Environment (IGU-CHE) and of the Young Scientists Seminar of the Lisbon Academy of Sciences. His current research focuses on the impact of socioeconomic determinants on vulnerability to environmental conditions.

Patrícia Pereira

Patrícia Pereira is a researcher at the Polytechnic of Leiria affiliated with the Interdisciplinary Center of Social Sciences (CICS.NOVA). She holds a Ph.D. in Sociology, and her main field of research and teaching is Urban Sociology. Her current work is focused on subjective experiences of eviction and displacement, while also considering the neighbourhood effects and structural dimensions of gentrification and, more broadly, of urban inequality. She specialises in qualitative methods, including ethnography and in-depth interviews, with a growing interest in creative methods and related ethical and knowledge production issues. She was recently elected to serve as Executive Committee member of the European Sociological Association.

land

MDPI

Editorial

Reimagining Urban Futures: Perspectives on Cultural, Environmental and Digital Transitions

Pedro Chamusca [1,*], André Carmo [2], Ricardo Almendra [3] and Patrícia Pereira [4]

1 Faculty of Arts and Humanities of the University of Porto, Via Panorâmica s/n, 4150-564 Porto, Portugal
2 Interdisciplinary Centre of Social Sciences, School of Sciences and Technology, University of Évora, 7004-516 Évora, Portugal; acarmo@uevora.pt
3 Centre of Studies in Geography and Spatial Planning, University of Coimbra, 3004-530 Coimbra, Portugal; ricardoalmendra85@gmail.com
4 CICS.NOVA, ESECS, Polytechnic of Leiria, 2411-901 Leiria, Portugal; ana.f.pereira@ipleiria.pt
* Correspondence: pedrochamusca@hotmail.com

Cities have become central arenas in confronting today's most pressing environmental and societal challenges, including climate change, biodiversity loss, socio-territorial inequalities and the disruptive impacts of technological innovation. With over 55% of the world's population now living in urban areas (with a projected 68% by 2050, according to the United Nations), cities play dynamic socio-ecological systems a defining role in shaping the well-being of both current and future generations. This reprint, entitled "Urban Transitions and Dynamics: Cultural, Ecological, Digital and Degrowth", emerged from growing academic and policy-driven demands to investigate cities as laboratories of transition, innovation and justice, as well as the need to explore how urban territories are being reconfigured in light of new environmental, technological and socio-political imperatives.

In recent years, we have witnessed a proliferation of concepts seeking to capture the new trajectories for urban sustainability. Notably, notions such as the 15-min city [1,2], circular urban economies [3,4], digital transitions [5,6], urban sustainable development [7,8], degrowth actions [9,10], and positive energy districts [11,12] have gained traction in both academic and planning discourse. Meanwhile, the implementation of major infrastructures—whether high-capacity transport corridors, intermodal complexes or "anchor" museums—has provoked ambivalent urban transformations. On the one hand, such projects stimulate local economies and attract fresh investment; on the other, they can exacerbate socio-spatial inequality by displacing vulnerable populations in the face of rising property values and housing costs. This dynamic intersects with touristification, which, although often driving the early stages of rundown neighbourhood rehabilitation and heritage preservation, ultimately generates mounting tensions between residents and visitors. This places immense pressure on the accommodation market, inflates the cost of living and gives rise to the "gourmet aestheticisation" of public spaces, turning squares and streets into segmented consumer venues. The gradual privatisation of these spaces—often justified in the name of security or visitor comfort—further commodifies everyday urban life, undermining social sustainability and the right to the city.

These frameworks share the ambition of reframing urban development around equity, low-carbon infrastructures and collective well-being. At the same time, critical voices have problematised the operationalisation of these paradigms, highlighting the risks of exclusion, technological determinism or reproducing old spatial injustices under new guises [13,14]. Moreover, there are growing concerns about the instrumentalization of these sustainability paradigms, whereby concepts originally rooted in social justice and environmental transformation are co-opted by market-driven actors or used as legitimation strategies for continued capital accumulation. This includes risks of greenwashing urban developments that maintain extractive logics under sustainable rhetoric or of social washing initiatives that claim community participation while reinforcing existing power structures, lifestyle brands or market niches that serve affluent populations while marginalizing vulnerable communities [15]. Such appropriations hollow out the transformative potential of these frameworks, converting visions for justice into instruments of capitalist accumulation.

Taken together, these emerging frameworks offer a rich analytical vocabulary and an ambitious normative horizon for guiding contemporary urban policy. The 15-min city, for instance, reconfigures the relationship between time, space and everyday life, promoting spatial equity through decentralising essential services [1,16]. Circular urban economies challenge the linear, extractive logic of conventional urban growth, encouraging regenerative cycles of production, consumption and waste management [10]. Meanwhile, degrowth perspectives introduce a necessary critique of the unsustainable foundations of urban development, inviting a revaluation of sufficiency, care, and non-market values in urban planning [17].

While enabling greater connectivity, data-driven management and new service models, digital transitions also raise crucial concerns regarding surveillance, digital divides and algorithmic governance. Similarly, positive energy districts exemplify how technological innovation can be leveraged to achieve energy autonomy and carbon neutrality at the neighbourhood scale; however, their implementation often requires new forms of institutional cooperation and citizen engagement. The success of these models thus hinges not only on technological feasibility but on their capacity to be embedded within inclusive and democratic planning processes.

In addressing the multifaceted crises cities face today, these concepts are not mutually exclusive but rather complementary. Their relevance lies in their potential to reshape planning paradigms and inform integrated strategies capable of navigating socio-environmental complexity. For planners, policymakers and researchers alike, the challenge is to interpret and adapt these ideas to diverse territorial realities, ensuring that they serve as tools for emancipation rather than instruments of exclusion or commodification.

The relevance of this reprint lies in its ability to bridge empirical insights and theoretical reflections on contemporary urban transformations. The nine chapters included explore diverse geographies and analytical lenses, engaging with issues such as green productivity, environmental well-being, digital transitions, new urban governance mechanisms, and the impacts of tourism and cultural dynamics on city centres. Together, they contribute to ongoing debates regarding how to foster more just, sustainable and resilient cities through integrated cultural, environmental, digital and post-growth approaches.

This endeavour was motivated by the recognition that the current wave of urban transitions demands a rethinking of spatial planning and territorial governance. As Guest Editors, our interests were both academic and political: to offer a platform for interdisciplinary dialogue and to inform better public policies that respond to territorial specificities while fostering urban sustainability and social cohesion. The diversity of cases and methods represented—from evaluating the impacts of high-speed rail to analysing in situ urbanisation through digitalisation—attests to the richness of this field, as well as the need for plural, context-sensitive understandings of urban change.

The nine contributions can be grouped into three thematic clusters. The first focuses on environmental transitions and well-being and includes papers by Zheng et al., Zhang & Zhou, and Hu et al. They propose new frameworks for integrating environmental indicators and green innovation into urban planning. The second thematic cluster addresses digital transitions and economic resilience, with contributions from Shi & Lu, Li et al. and Li

& Song exploring how digitalisation reshapes urban systems, labour geographies and economic adaptability. The final theme includes place-based studies that examine cultural, infrastructural and tourism-led transformations in urban centres and includes works by Fernandes et al., Tian et al., and Chi & Han.

This reprint is primarily aimed at academic audiences working in the fields of geography, urban and environmental engineering, architecture, sociology, development studies, political economy and related disciplines. Its interdisciplinary scope and methodological diversity are especially relevant for scholars analysing the interplay between urban form, socio-technical systems, governance mechanisms and sustainability transitions. The collected articles provide both conceptual depth and empirical substance, making them valuable resources for those researching urban dynamics and spatial justice in an age of environmental urgency and technological change.

At the same time, this volume is intended to inform the work of practitioners and decision-makers involved in urban planning and governance. It is particularly directed at municipal planners, technical staff and local government leaders who face the daily challenge of implementing sustainable policies in diverse and often complex urban contexts. By offering grounded insights, comparative examples and critical perspectives, this reprint aims to support the formulation of more effective, equitable and context-sensitive public policies.

We invite readers to engage critically and constructively with the contributions presented here. The pathways towards more sustainable and inclusive urban futures are plural and contested, but they require collective imagination and empirical grounding. We hope that this Special Issue provides inspiration, tools and prompts for researchers, planners, policymakers and citizens committed to rethinking urban life in the face of environmental and social emergencies.

We wish to thank all the collaborative reviewers who contributed to improving the chapters in this work.

Author Contributions: Conceptualization, P.C., A.C., R.A. and P.P.; formal analysis, P.C.; investigation, P.C., A.C., R.A. and P.P.; resources, P.C., A.C., R.A. and P.P.; writing—original draft preparation, P.C.; writing—review and editing, P.C., A.C., R.A. and P.P.; visualization, P.C., A.C., R.A. and P.P.; supervision, P.C., A.C., R.A. and P.P.; project administration, P.C., A.C., R.A. and P.P. All authors have read and agreed to the published version of the manuscript.

Acknowledgments: We express our gratitude to all the authors, reviewers and editorial staff that worked on this Special Issue.

Conflicts of Interest: The authors declare no conflicts of interest.

References

1. Moreno, C.; Allam, Z.; Chabaud, D.; Gall, C.; Pratlong, F. Introducing the "15-minute city": Sustainability, resilience and place identity in future post-pandemic cities. *Smart Cities* **2020**, *4*, 93–111. [CrossRef]
2. Rao, F.; Kong, Y.; Ng, K.H.; Xie, Q.; Zhu, Y. Unravelling the Spatial Arrangement of the 15-Minute City: A Comparative Study of Shanghai, Melbourne, and Portland. *Plan. Theory Pract.* **2024**, *25*, 184–206. [CrossRef]
3. Kębłowski, W.; Lambert, D.; Bassens, D. Circular economy and the city: An urban political economy agenda. *Cult. Organ.* **2020**, *26*, 142–158. [CrossRef]
4. Lakatos, E.S.; Yong, G.; Szilagyi, A.; Clinci, D.S.; Georgescu, L.; Iticescu, C.; Cioca, L.-I. Conceptualizing Core Aspects on Circular Economy in Cities. *Sustainability* **2021**, *13*, 7549. [CrossRef]
5. Daniel, C.; Pettit, C. Digital disruption and planning–use of data and digital technology by professional planners, and perceptions of change to planning work. *Aust. Plan.* **2021**, *57*, 50–64. [CrossRef]
6. Potts, R.; Milz, D. A culture of digital planning? An international comparison of culture, planning and technology. *Int. Plan. Stud.* **2024**, *29*, 327–341. [CrossRef]

7. Praharaj, S. Area based urban renewal approach for smart cities development in India: Challenges of inclusion and sustainability. *Urban Plan.* **2021**, *6*, 202–215. [CrossRef]
8. Wei, G.; He, B.-J.; Sun, P.; Liu, Y.; Li, R.; Ouyang, X.; Luo, K.; Li, S. Evolutionary trends of urban expansion and its sustainable development: Evidence from 80 representative cities in the Belt and Road Initiative region. *Cities* **2023**, *138*, 104353. [CrossRef]
9. Kaika, M.; Varvarousis, A.; Demaria, F.; March, H. Urbanizing degrowth: Five steps towards a Radical Spatial Degrowth Agenda for planning in the face of climate emergency. *Urban Stud.* **2023**, *60*, 1191–1211. [CrossRef]
10. Savini, F. Strategic planning for degrowth: What, who, how. *Plan. Theory* **2024**, *24*, 141–162. [CrossRef] [PubMed]
11. Battaglia, V.; Massarotti, N.; Vanoli, L. Urban regeneration plans: Bridging the gap between planning and design energy districts. *Energy* **2022**, *254*, 124239. [CrossRef]
12. Martinelli, V. Energy transition: Digital (t)win? *TeMA—J. Land Use Mobil. Environ.* **2023**, *16*, 653–659. [CrossRef]
13. Marcuse, P. From critical urban theory to the right to the city. *City* **2009**, *13*, 185–197. [CrossRef]
14. Soja, E.W. *Seeking Spatial Justice*; University of Minnesota Press: Minneapolis, MN, USA, 2010; ISBN 9780816666683.
15. Anguelovski, I.; Connolly, J.J.T.; Pearsall, H.; Shokry, G.; Checker, M.; Maantay, J.; Gould, K.; Lewis, T.; Maroko, A.; Roberts, J.T. Why green "climate gentrification" threatens poor and vulnerable populations. *Proc. Natl. Acad. Sci. USA* **2019**, *116*, 26139–26143. [CrossRef]
16. Pirmoradi, P.; Garshasbi, D.; Solgi, N. A critical review: The 15-minute city. *Plan. Pract. Res.* **2025**, *40*, 1–22. [CrossRef]
17. Iveson, K. Social or spatial justice? Marcuse and Soja on the right to the city. *City* **2011**, *15*, 250–259. [CrossRef]

land

MDPI

Article

Unlocking Economic Resilience: A New Methodological Approach and Empirical Examination under Digital Transformation

Chenchen Shi [1,2,*] and Jinjing Lu [1]

[1] School of Urban Economics and Public Administration, Capital University of Economics and Business, Beijing 100070, China; lujinjing325@163.com

[2] Beijing Key Laboratory of Megaregions Sustainable Development Modeling, Capital University of Economics and Business, Beijing 100070, China

* Correspondence: ccshi@cueb.edu.cn

Abstract: Economic resilience is crucial for urban sustainability as it ensures stability and growth in the face of external shocks, promotes social cohesion and inclusivity, fosters environmental sustainability, and enhances cities' adaptability to future challenges. This study expands the conventional perspective on economic resilience beyond the context of shocks, focusing on the inherent resilience of regional economic systems. A novel method for quantifying economic resilience is introduced, emphasizing system sensitivity and adaptability. Using Chinese prefecture-level city data and an econometric model, we empirically examine how Fintech, a major digital transition in current urban systems, affects economic resilience. The findings reveal that Fintech has a substantial positive effect on economic resilience, primarily through the upgrading of industrial structures and technological innovation. Furthermore, there is significant regional heterogeneity in the impact of Fintech on economic resilience, with more pronounced contributions in the east, central, and western regions of China, as opposed to the northeast. Additionally, the impact of Fintech on economic resilience is more substantial in large-scale cities. The promotion of economic resilience through digital transformation serves as a potent risk prevention measure. Understanding the role of economic resilience in urban systems holds valuable implications for countries worldwide.

Keywords: economic resilience; urban systems; digital transition; regional heterogeneity; China

1. Introduction

Urban sustainability encompasses the ability of a city to maintain economic, social, and environmental well-being while meeting the needs of its current and future residents. In urban systems, the development trajectories of regional economies are constantly evolving and subject to various shocks. To examine the resilience of regional economic systems to these shocks, scholars have drawn upon system equilibrium models from the fields of physics and ecology [1–3] and have proposed the concept of "economic resilience", which builds upon the principles of engineering and ecological resilience [4]. This concept has gained traction due to its ability to capture the holistic and dynamic nature of socioeconomic systems and its potential to inform policy and academic research [5,6]. Resilience is considered an important expression of sustainability and is vividly referred to as the emergency room of sustainability [7]. The importance of economic resilience for urban sustainability lies in its ability to maintain stability and foster growth amidst external disturbances, while also bolstering social unity and inclusivity, promoting environmental sustainability, and strengthening cities' capacity to adapt to future challenges [5,8]. Recent years have witnessed heightened global uncertainty, stemming from events such as the 2008 financial crisis, trade tensions between China and the United States, the COVID-19 pandemic, and wars and conflicts, as well as other shocks that have generated ongoing economic turmoil. The ability of regions to withstand uncertainties in the economic system has become an increasingly pressing concern for governments and academics alike [9,10].

Financial development is a crucial driver of economic growth and prosperity. During the digital transition, the rise of financial technology (Fintech)[1], which integrates technology and financial services, provides new opportunities for the development of the traditional financial system [12]. Digital transactions powered by Fintech optimize financial resource allocation, drive technological innovation, and facilitate the transformation and upgrading of financial infrastructure [13]. This integration enhances financial efficiency, inclusivity, and innovation, satisfying the personalized financial needs of diverse businesses [14]. According to the World Bank, global Fintech investment soared from less than $10 billion annually before 2013 to $215 billion in 2019 and then decreased to $122 billion in the 2020 pandemic year. By the first half of 2021, global investments in Fintech had reached $98 billion [15]. Although existing studies have explored the impact of financial development on economic resilience [16] and the influence of regional technological structures [17,18], as well as the effects of smart city construction, including its digital transitions, on economic resilience [19], research specifically examining the direct impact of Fintech on economic resilience remains limited [20,21].

This study aims to examine economic resilience within the context of digital transformation in the financial sector. While past research has predominantly focused on economic resilience in response to shocks, we aim to broaden the horizon by examining the inherent resilience of regional economies and its relationship with Fintech innovation. In an era where Fintech is transforming traditional financial systems and playing an increasingly pivotal role in economic development, understanding its direct influence on economic resilience, regional variations, and its underlying mechanisms becomes not only relevant but also crucial. This understanding has the potential to provide valuable insights for policy decisions and strategic investments in a rapidly evolving urban and technological landscape.

To address this research aim, this study commences by constructing a quantification method to assess regional economic resilience. Subsequently, it empirically examines the impact of Fintech on economic resilience using data from Chinese prefecture-level cities during the period from 2011 to 2020 (Figure 1). The research delves into the underlying mechanisms, taking into consideration aspects like technological innovation and industrial structure upgrading, to explain this impact. Additionally, it employs cross-sectional analysis to accommodate the variations arising from geographical locations and city scales. The reasons for selecting China as the empirical case study region are as follows. The diverse geographic and economic landscape of China provides a rich and varied context for analyzing how different regions leverage technological innovation and industrial upgrades to enhance their economic resilience. In addition, China's rapid digital transformation, especially in the financial sector, with widespread adoption of Fintech solutions like mobile payments and digital banking, presents a unique opportunity to explore how such technologies influence economic stability. This setting allows for a detailed examination of the dynamic interplay between digital transition and economic resilience across varying urban scales and geographical locations.

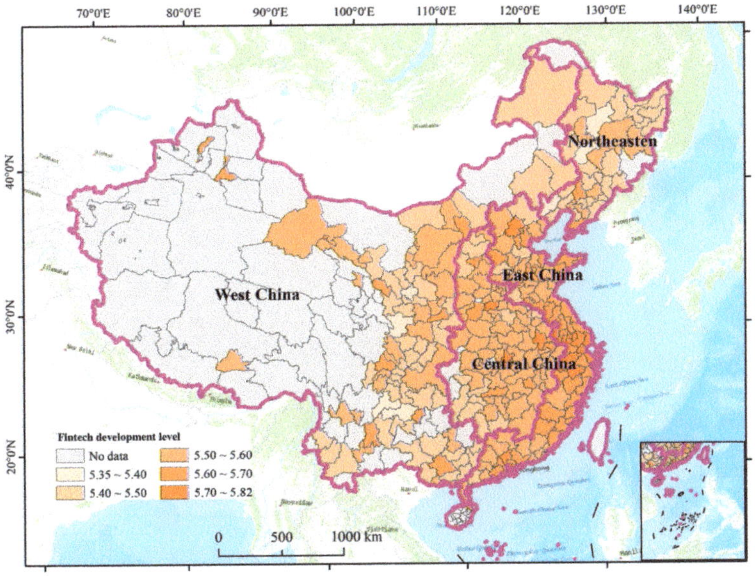

Figure 1. Location and Fintech development of sample cities in China (2020).

2. Literature Review and Conceptual Development

2.1. Economic Resilience and Its Operationalization

2.1.1. The Notion of Economic Resilience

The concept of economic resilience, as it is understood in the context of regional economics and economic geography, does not have a singular point of origin but rather evolved over time. It emerged from the broader discussion of resilience in various disciplines, including ecology, psychology, and engineering [4]. The term began to gain more prominence in economic discussions particularly following global economic challenges, such as financial crises, which underscored the need to understand how economies can withstand and recover from shocks [22,23].

Over the years, the conceptualization of economic resilience in regional contexts has evolved into a multifaceted and dynamic discourse, highlighting the complex nature of regional economic resilience and its multi-dimensional and adaptive qualities. Boschma [24] and Bristow and Healy [25] underscore the adaptive nature of regional economies, arguing that resilience involves not just "bouncing back" but also "bouncing forward" through adaptation and transformation in response to shocks. Hu and Hassink [26] further argue that adaptation, seen as a continuous, evolving process, plays a pivotal role in how regions respond to economic challenges and disturbances. These perspectives align with the evolutionary approach to resilience, which sees regional economies as dynamic systems constantly adapting to their changing economic landscapes.

Evenhuis [27], Martin and Sunley [28], and Sutton and Arku [29,30] contribute by examining the dimensions of regional economic resilience. They emphasize the importance of regional specificities and the role of external and internal factors, such as policy, institutional settings, and socio-economic structures, in shaping a region's resilience. Their work highlights that resilience is not a static attribute but a process that evolves over time, influenced by a region's unique characteristics and its ability to adapt to changing circumstances.

The previous research provides a comprehensive view of regional economic resilience, framing it as a dynamic, adaptive, and multi-dimensional concept that varies across regions and over time. The emphasis on adaptation and transformation, as well as the recognition of the diverse factors that influence resilience, marks a significant evolution in the understanding of economic resilience in regional contexts. While some scholars suggest that regional economic resilience should only be examined in the context of shocks [5], this study conceptualizes resilience as the ability to sense and adapt to both shocks and slow-burning pressures [31] and places its empirical research within a broader context.

2.1.2. Evolutionary and Equilibrium Perspective of Resilience

In the literature on regional economic resilience, two distinct perspectives are highlighted: the equilibrium and the evolutionary approaches [29]. The equilibrium approach assumes that economies naturally gravitate towards a steady state, and following a disturbance, they can return to or reach a new equilibrium [32–34]. The evolutionary approach, in contrast, posits that economies are constantly adapting, without an inherent steady state, and resilience involves their capacity to evolve in the face of shocks [22,25,35].

This study aligns with the equilibrium perspective, which can be justified according to its focus on measurable outcomes, such as GDP, allowing for the assessment of resilience as a system's return to its pre-shock state or its shift to a new stable state after a disruption. This perspective provides a structured framework for analysis, making it possible to evaluate economic performance systematically. A distinctive feature of this study is the conceptualization of resilience as an intrinsic attribute of a system, allowing for the examination of resilience under normal conditions, not just in the aftermath of shocks. This approach facilitates a proactive understanding of resilience and enables the development of strategies to strengthen economic systems in preparation for potential disruptions.

Due to the varying conceptualization of economic resilience among scholars discussed before, the methods of operationalizing economic resilience also differ[2]. Two common quantitative approaches to characterizing economic resilience are using the sensitivity index developed by Martin et al. [3,23,38,39] and the development of a comprehensive indicator system to measure economic resilience [40–42]. However, the sensitivity index measures the resilience of a single city (system) based on its spatial divergence from other cities (systems), ignoring temporal differences. And the comprehensive indicator system primarily evaluates resilience based on static attributes, disregarding the dynamic nature of the concept of resilience itself and the adaptability of the assessed system. Accordingly, this study proposes constructing a new economic resilience measurement method.

2.1.3. Sources/Determinants of Economic Resilience

The determinants of resilience within regional economies are complex, multifaceted, and crucial for understanding why some regions recover from shocks more quickly or effectively than others. Martin et al. [3] elaborate on how regional economic structures, the diversity and quality of resources, and capabilities, along with institutional support, are crucial determinants that shape the capacity of regions to withstand and recover from economic shocks. Evenhuis [27] adds to the dialogue by emphasizing adaptability—the capability of regional economic systems to maintain and restore their structure in the face of disturbances, which acts as a self-regulating mechanism for maintaining stability.

Sutton and Arku [30] further explore the multi-scalar interactions of various factors within regions, including the role of economic actors, policymakers, and practitioners, and how these interactions influence the resilience of a region's economic structure and workforce. Sutton et al. [29] identify the determinants of resilience as being dynamic, multi-scalar, and spatially dependent, shaped by a wide range of socio-economic and political-institutional factors that are influenced by regions' inherent and inherited resources, capabilities, and characteristics. These factors, such as human capital, agglomerations, entrepreneurship, and innovativeness, not only bolster regions' resilience in turbulent times but also tend to enhance their growth potential and competitiveness during stable periods.

This study, set within the burgeoning context of China's digital transformation in the financial technology sector, seeks to uncover new sources of economic resilience within financial arrangements. It asserts that financial arrangements can be a significant source of resilience (in alignment with Martin et al., 2016 [3]), offering novel insights into how economic systems can leverage financial innovations and structures to enhance their stability and adaptability in the face of economic perturbations. This way of conceptualizing resilience—not merely as a response to adverse events but as an inherent attribute of the economic system itself—is one of the distinctive features of this study. By focusing on financial arrangements, this study extends the discourse on economic resilience, providing a unique angle on how digital transitions contribute to the resilience of regional economies.

2.2. *Fintech and Economic Resilience*

Financial development is a pivotal driver of sustainable economic growth, exerting its influence across the economic spectrum by facilitating fund allocation. Consequently, it plays a crucial role in bolstering economic resilience. Furthermore, the fusion of financial development with digital technology has spawned Fintech, heralding both fresh prospects and hurdles in fortifying economic resilience [43]. This synergistic blend holds promise in propelling sustainable economic expansion and bolstering the adaptive and recovery capabilities of regional economies amidst shocks and disruptions.

2.2.1. Fintech's Impact on Economic Resilience

The rise of Fintech has transformed traditional finance, introducing innovative services, scenarios, and models. These advancements bridge the gap between financial supply and demand, reducing mismatches and expanding the financing options for businesses. Consequently, financing costs decrease, profitability increases, and enterprise sustainability improves [44,45]. Fintech's technological efficiency accelerates fund matching, aiding rapid enterprise recovery and adaptation during financial crises [46].

Moreover, Fintech's global reach extends financial services to underdeveloped regions via digital platforms, enhancing inclusivity, streamlining financing, and improving capital turnover, thus enhancing resource allocation efficiency [12,47,48]. Fintech's role in reducing the information asymmetry between institutions and markets enhances risk management, crucial for economic resilience during shocks.

Fintech's technology and knowledge transfer stimulate structural optimization and industrial advancement, fostering cross-regional flows and scalability, essential for market resilience [49]. By facilitating swift adjustments to economic shifts, Fintech enhances enterprises' adaptive capacity, supporting strategic realignment amidst changing landscapes. This emphasis on adaptability underscores Fintech's transformative role, positioning it as a fundamental enabler of resilient economic structures capable of thriving through financial crises and downturns.

Based on the above analysis, we propose the following hypothesis in the empirical investigation of this study:

H1: *Fintech development has a positive impact on economic resilience.*

2.2.2. The Influence Mechanism

The outputs of science, technology, and innovation (STI) often entail high inputs, lengthy cycles, and considerable uncertainty, posing challenges in accessing financing within traditional financial systems. Fintech, as a disruptive financial innovation [48], is bolstering technological innovation behaviors among market participants, enhancing financial market efficiency, and reducing operational costs while identifying prospects and risks through advanced technologies [50]. By tailoring financial services to specific market needs, Fintech improves risk management and recovery, thereby enhancing economic resilience.

Furthermore, Fintech facilitates the timely and accurate delivery of STI project information to financial institutions, reducing information asymmetry and enabling flexible access to financial services for market players [51]. This, in turn, fosters technological innovation activities, promoting the development of science and technology industries and fortifying economic resilience. Additionally, in line with evolutionary economics, firms utilize technological innovation to respond to external market shocks, revitalizing economic ecosystems and forging new development pathways [52]. Hence, we posit the hypothesis:

H2: *Fintech can enhance economic resilience by improving technological innovation.*

Moreover, Fintech's impact extends beyond technological innovation to optimize industrial structures [50]. By diversifying financing channels and facilitating efficient resource allocation, Fintech propels the transformation and upgrading of traditional industries, particularly in China, where incentives for structural innovation may be lacking [53]. Through enhanced factor mobility, information dissemination, and coordination within industrial chains, Fintech promotes industrial restructuring, bolstering the economy's resilience to risks. Additionally, the development of financial technology spurs the emergence of new models and industries, elevating the industrial structure to higher levels. The spillover effects of digital financial services provide enterprises with superior credit services and stimulate employment growth, particularly in burgeoning service and production sectors. Thus, continuous industrial restructuring yields high value-added outputs, enhancing "structural dividends" and propelling both economic growth and quality development [54]. Building on these insights, we propose the following hypothesis:

H3: *Fintech can enhance economic resilience by improving the industrial structure.*

The conceptual framework in Figure 2 illustrates Fintech's direct impact on regional economic resilience (H1), accounting for regional disparities and city-scale differences. Additionally, the study explores its indirect effects through the mediating roles of technological innovation (H2) and industrial structure (H3).

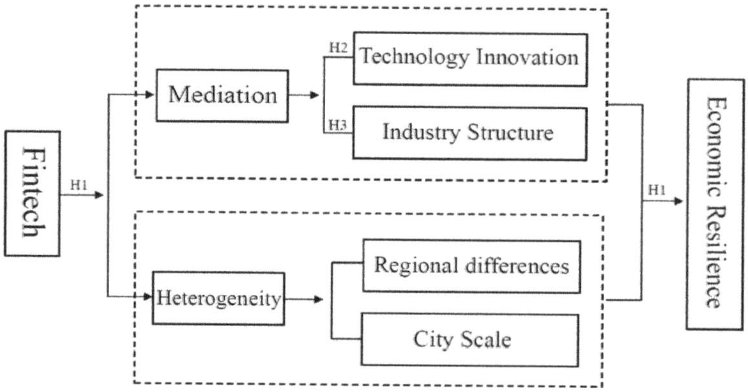

Figure 2. Conceptual framework of Fintech's impact on economic resilience.

3. Economic Resilience Measurement Method

3.1. Selection of Economic Resilience Indicators

In the previous research, the two most crucial indicators for measuring economic resilience have been identified as GDP and employment. While data on employment offer valuable insights at the individual level [55], GDP serves as a comprehensive measure of a city's overall economic performance. This broader perspective is vital for fully understanding the resilience of urban economies, capturing not only employment status but also the economic activities and outputs that sustain the livelihoods of the city's inhabitants.

Specifically, GDP analysis presents an appropriate means of measuring economic resilience by scrutinizing the capacity of an economy to withstand shocks. This involves examining changes in GDP growth rates over time and comparing them with pre-shock levels. In this study, we used the GDP indicator to measure regional economic resilience and construct an economic resilience index, which will be elaborated in the following sub-section.

3.2. Construction of an Economic Resilience Index

3.2.1. Sensitivity

Sensitivity refers to how responsive a system is to disruptions during its regular operation. It is a relative index that measures regional/local economies' performance compared to the national performance [3,23]. This study employs GDP to represent the functionality of the economic system, and system sensitivity is measured by analyzing the year-to-year changes in the GDP, reflecting the extent of deviation of each year's GDP from the average value across the years within the system. The formula for calculation is expressed as follows:

$$S_{it} = \frac{|P_{it} - \overline{P_i}|}{\overline{P_i}} \tag{1}$$

where i is the region, and P_{it} is the GDP value in region i in year t. $\overline{P}_i$ is the mean value of the GDP in region i across years. S is the sensitivity index, which reflects the dispersion of the GDP relative to the mean value over a specific period.

To demonstrate sensitivity more visually, we plotted Figure 3 using data from the empirical case of 286 Chinese cities, where the x axis is the sample city ranked from the lowest average GDP to the highest average GDP from left to right and the y axis is the GDP value. The red line in the figure represents $\overline{P}_i$, the mean value of the GDP in each city from 2011 to 2020. And the blue dot represents the GDP value of each city in 2016, i.e., P_{it}. The vertical distance from the blue dot to the red line is the numerator of Equation (1).

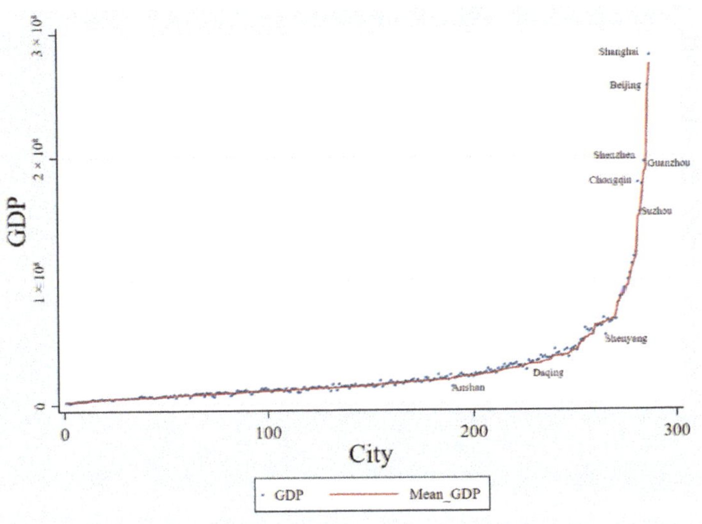

Figure 3. Sensitivity index evaluation results of the empirical case.

3.2.2. Adaptability

Adaptability refers to the ability of an economic system to maintain and restore its structure in the face of disturbances [26]. It can also be seen as a self-regulating mechanism for maintaining the system's relatively stable state. Within a certain period, the trend in the variability in a system is used to measure its deviation from the stable state, which is the system's adaptation. If the trend in its variability decreases or remains stable, the system tends towards relative stability. Increased variability indicates that an unstable system is adapting to changes, which may also indicate an increase in vulnerability.

Over a certain period, adaptability, which is the self-regulating capacity of the system, can be expressed as the slope of a linear trend line fitting the interannual variability in the GDP. In this study, adaptability is represented by the slope of a linear trend line fitting the interannual variability in the GDP during the study period.

$$y = A_i x + B_i \tag{2}$$

$$y = P_{it} - \overline{P_i} \tag{3}$$

where x represents the time series, corresponding to the study period; y represents the interannual variability in the GDP, which is the absolute change in the GDP each year, calculated as the difference between the value of the GDP for each year and the average value of the GDP during the study period. A represents the trend in the GDP's variability, which is the regression slope of the data set y and x and serves as the index of system adaptability. B represents the intercept.

For a more visual representation, we have selected three Chinese cities, Shanghai, Beijing, and Shenzhen, and plotted Figure 4 to show the fit of adaptation in these three cities, where the x axis is the year and the y axis is the calculation results of Equation (3). The blue dot is the interannual variability in the GDP, i.e., the y in Equations (2) and (3). And the red line is the linear fitting of the data set y and x. Thus, the A index is the slope of the red line.

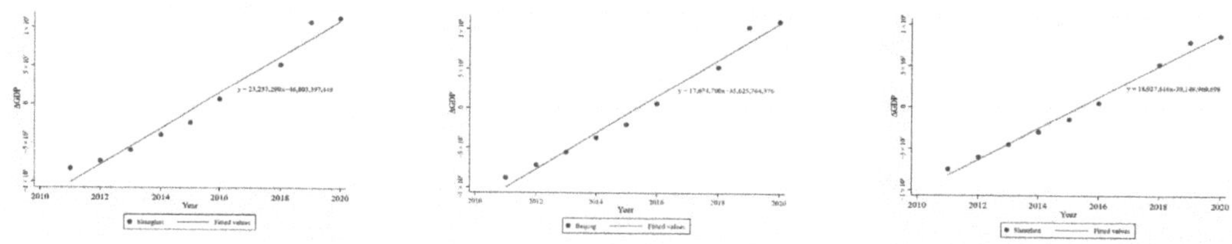

Figure 4. Adaptability index evaluation results of the empirical case. Note: The year 2017 is excluded with no data in the year 2017 from the original data source.

3.2.3. Economic Resilience

Resilience is determined by the system's sensitivity and adaptability, which is inversely correlated with sensitivity and positively correlated with adaptability. To compare regional differences in resilience, the results of sensitivity and adaptability should be standardized before calculating resilience. The formula for calculating resilience is as follows:

$$R = A - S \tag{4}$$

where R is system resilience, S is the sensitivity index, A is the adaptability index.

The proposed methodology refines and advances existing frameworks by presenting a sophisticated approach to the measurement of economic resilience. This approach diverges from the models suggested by Martin et al. [3,23], which primarily evaluate regional performance relative to national benchmarks. Instead, it delves into a comprehensive examination of regional variability and stability via GDP temporal variation analysis within one single economy. Furthermore, while this study adopts an equilibrium perspective for the conceptualization and quantification of resilience, the economic resilience index developed herein adeptly encapsulates the sensitivity and adaptability inherent to economic systems. This index portrays resilience not merely as a reactive capacity but as an inherent attribute of the system, enabling it to navigate and adapt to both immediate disruptions and ongoing changes effectively.

4. Empirical Analysis

4.1. Economic Resilience of Chinese Cities

To empirically apply the economic resilience index, we conducted an analysis using data from 286 Chinese cities for the years 2011, 2014, 2016, and 2020[3]. The ArcGIS platform was employed to visualize the assessment results. The economic resilience results have been divided into five levels: high (>1.5), mid-high (0.5~1.5), medium (−0.5~0.5), mid-low (−1.5~−0.5), and low (<−1.5) (Figure 5).

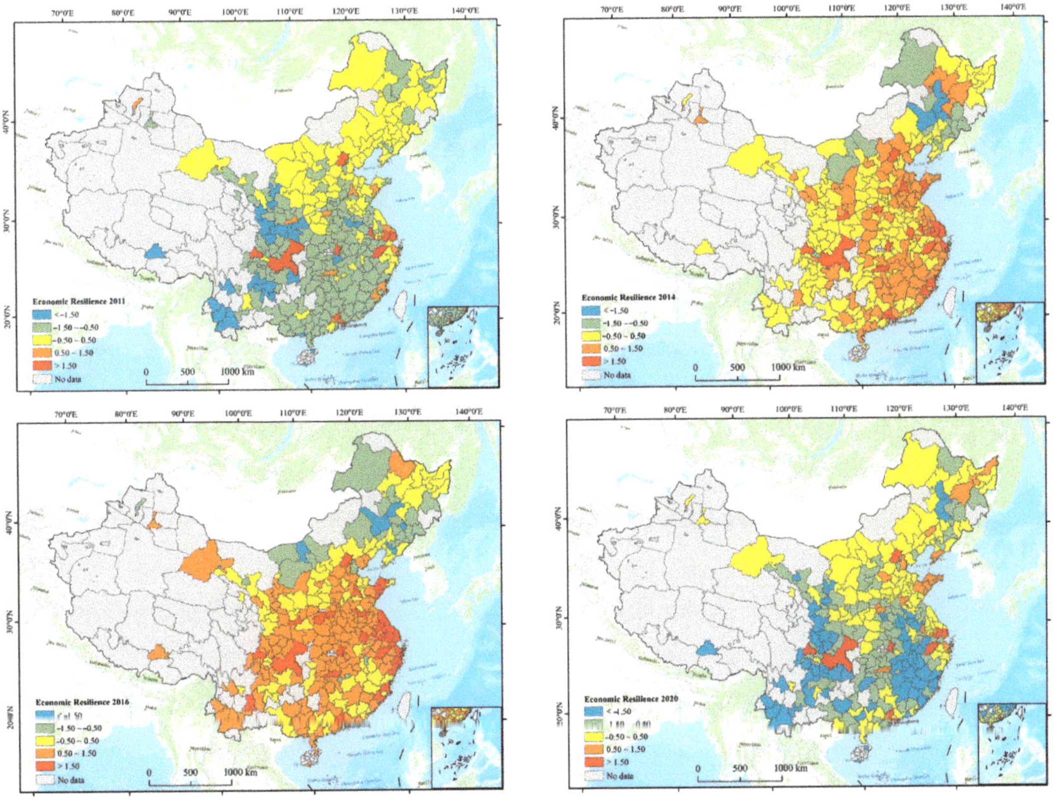

Figure 5. Economic resilience of 286 prefecture-level cities in China from 2011 to 2020.

From 2011 to 2016, economic resilience in Chinese cities is generally increasing. Meanwhile, the progressive expansion of red–orange areas (high and mid-high resilience) until 2020 suggests a significant shock to the economic resilience of each prefecture-level city in China, possibly due to the unforeseen black swan event of the Covid pandemic. In terms of spatial heterogeneity, though the study period, cities with a relatively higher level of economic resilience are mainly concentrated major urban agglomerations like the Beijing–Tianjin–Hebei region, the Yangtze River Delta region, and the Pearl River Delta region. In addition, the findings of our study also corroborate prior research indicating that regional economic resilience in China exhibits spatial heterogeneity, with municipalities and provincial capitals tending to display higher levels of resilience [56,57]. The evaluation results contribute to a deeper understanding of the spatio-temporal evolution of economic resilience in China and may inform future policy decisions.

4.2. Relationship between Fintech and Economic Resilience

Following the comprehensive analysis of the spatio-temporal variations in economic resilience across cities, the focus of this study shifts towards exploring the intricate relationship between Fintech and economic resilience. This transition marks a critical juncture in our research, bridging the gap between the foundational understanding of how cities withstand and adapt to economic fluctuations and the role of Fintech in enhancing or transforming these resilience capacities.

From 2011 to 2016, the relationship between Fintech and economic resilience grew stronger, but by 2020, this relationship had weakened (Figure 6). Despite this, the overall trend still indicates that enhancements in Fintech are positively correlated with higher economic resilience. However, other determinants of economic resilience became more significant in 2020, which may reflect the increasing influence of other economic and social factors alongside the rapid development of Fintech. Therefore, in the following sub-section, we will delve deeper into the dynamics of this relationship, incorporating additional control variables to provide a more nuanced analysis.

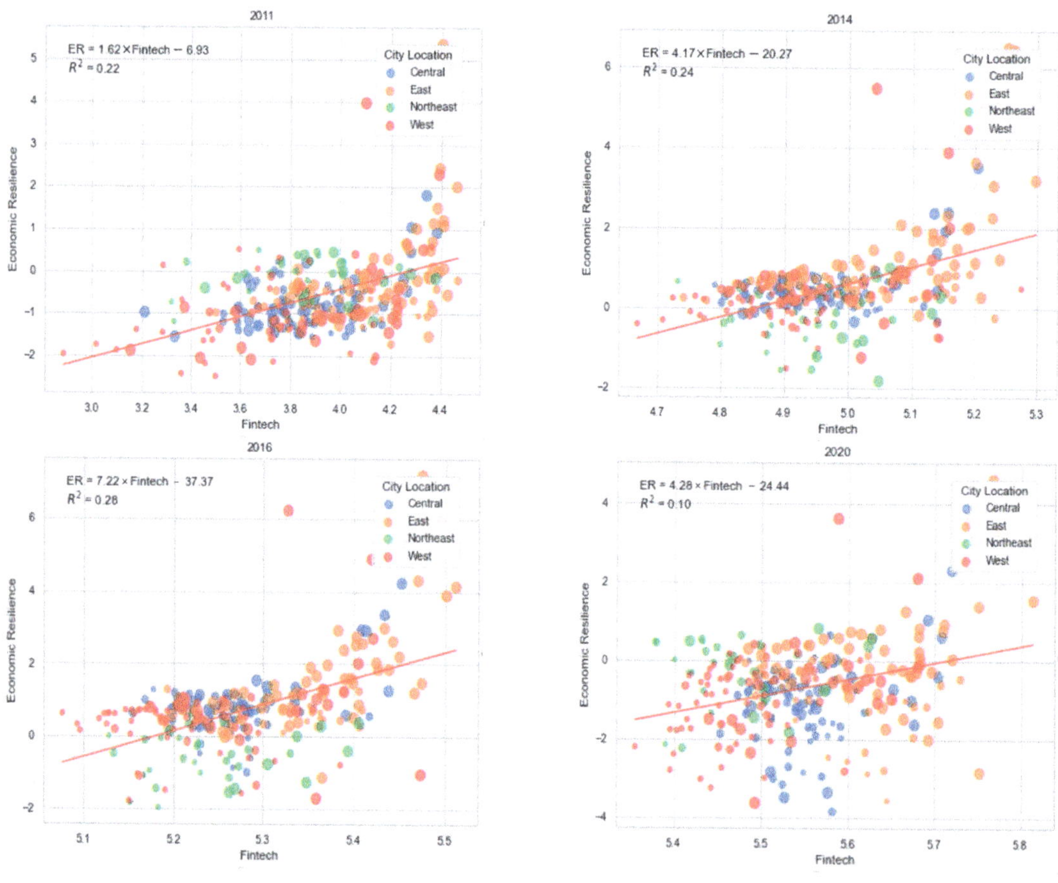

Figure 6. Analysis of the relationship between Fintech and economic resilience in Chinese cities from 2011 to 2020. **Note**: The size of the dots in the graph indicates the size of the city, while the color of the dots represents the city's location.

4.3. Finding New Sources of Economic Resilience: Fintech

4.3.1. Data Source and Description

To standardize the data, the sources for each indicator were primarily drawn from the China City Statistical Yearbook, the Digital Inclusive Finance Index of Peking University, and the China Statistical Yearbook. To investigate the impact of Fintech on economic resilience, we selected indicators relevant to Fintech and economic resilience across 286 prefecture-level cities in China from 2011 to 2020. Furthermore, to mitigate the effects of outliers on the study's findings, we conducted winsorization for continuous variables at the 1% level.

4.3.2. Variable Selection

The variables used in this study are shown in Table 1. *Explained variable*: Economic resilience (ER), measured based on the economic resilience index in Section 3. The assessment results are illustrated in Figure 5 in the previous sub-section. *Explanatory variable*: Financial technology (Fintech). In this research, referring to Zhou et al. [58] and Guo et al. [59], we select the total index of China's digital inclusive finance published by the Institute of Digital Finance at Peking University as the measure of Fintech[4], which can comprehensively reflect the progress of China's digital financial inclusion and the development level of Fintech in prefecture-level cities. *Control variables*: In this study, we identified four control variables, including financial development (FD), economic openness (Open), human capital (HC), and fiscal position (FS). By controlling for these variables, we can isolate the effect of the independent variable on economic resilience while accounting for the effects of other confounding factors.

Table 1. Variable definitions.

Variable Types	Variable Name	Variable Symbol	Variable Selection
Explained variable	Economic resilience index	ER	Self-constructed economic resilience index
Explanatory variables	Financial science and technology	Fintech	Digital Financial Inclusion Index
Control variables	Financial development	FD	(Year-end financial institution deposit balance + year-end financial institution loan balance)/GDP
	Economic openness	Open	(Goods imports million yuan + goods exports million yuan)/regional GDP
	Human capital	HC	The number of college students per 10,000 students is added by 1 and the natural logarithm is taken
	Financial status	FS	Local fiscal general budget expenditure/regional GDP

4.3.3. Econometric Model

Referring to the study by Mai et al. [39], we use a panel regression model to study the impact of Fintech on economic resilience. In the model, the economic resilience index (ER) is the explanatory variable, the level of Fintech is the explanatory variable, and CV is the control variable.

$$ER_{it} = \alpha_0 + \beta_1 Fintech_{it} + i\beta CV_{it} + \varepsilon_{it} \tag{5}$$

where i represents the 286 prefecture-level cities in China, and t represents the year. ε represents a random perturbation term.

5. Results and Discussion

5.1. Descriptive Statistics

The results on the descriptive statistics of the variables are shown in Table 2. Panel A provides a descriptive overview. The analysis reveals that economic resilience varies considerably across regions in China. This variation highlights the differences in the ability of regional economies to withstand risks and external shocks. Similarly, the Fintech variable exhibits substantial spatial imbalance. The large difference between the minimum and maximum values indicates that the level of Fintech is unevenly distributed across regions.

Panel B presents the evolutionary trends in economic resilience and Fintech from 2011 to 2020. The mean values show that economic resilience fluctuates over time, with a decline in 2020 due to the sudden outbreak of the epidemic in 2019, which negatively impacted the Chinese economy. In contrast, the Fintech variable displays a steady upward trend from 2011 to 2020.

Table 2. Descriptive statistics.

Panel A								
Variable	**Obs**	**Mean**	**SD**	**Min**	**P25**	**Median**	**P75**	**Max**
ER	2573	−0.004	1.234	−2.897	−0.735	−0.008	0.646	4.917
Fintech	2573	5.025	0.518	3.533	4.784	5.152	5.434	5.700
FD	2573	2.458	1.162	0.987	1.661	2.142	2.874	7.017
Open	2573	0.051	0.138	0.000	0.000	0.000	0.028	0.919
HC	2573	2.635	2.506	0.000	0.000	3.384	4.989	6.964
FS	2573	0.204	0.102	0.077	0.135	0.177	0.241	0.626

Panel B					
Variable	**Year**	**Min**	**Max**	**Mean**	**Standard error**
ER	2011	−2.897	4.704	−0.781	1.062
	2014	−1.818	4.917	0.625	0.908
	2016	−1.909	4.917	0.850	1.106
	2020	−2.897	4.053	−0.845	1.312
Fintech	2011	3.533	4.472	3.938	0.262
	2014	4.978	5.300	4.978	0.122
	2016	5.078	5.513	5.279	0.089
	2020	5.356	5.700	5.546	0.084

5.2. Benchmark Regression

We conducted benchmark regression using panel data from 286 prefecture-level cities in China to examine the relationship between Fintech and economic resilience. Our stepwise regression approach initially did not incorporate any relevant control variables. The results from Table 3, Column (1) demonstrate a significant enhancing effect of Fintech on economic resilience. Results in Column (2) of Table 3 suggest that the positive effect of Fintech on economic resilience remains unchanged after introducing control variables. Specifically, a 1% increase in the Fintech innovation level can lead to a 0.835% increase in economic resilience. These findings contribute to the growing body of literature on Fintech and its implications for economic development, particularly in the context of China's prefecture-level cities.

Table 3. Benchmark regression.

	(1)	**(2)**
	ER	ER
Fintech	0.412 ***	0.835 ***
	(9.25)	(17.08)
FD		0.230 ***
		(7.60)
Open		−0.354
		(−1.21)
HC		0.188 ***
		(15.26)
FS		−2.273 ***
		(−9.44)
_cons	−2.074 ***	−4.779 ***
	(−9.63)	(−19.08)
N	2573	2573
R^2	0.031	0.222

Note: Robust t-statistics in brackets. *** $p < 0.01$.

5.3. Endogenous Test

Current scholarly consensus suggests that Fintech development is positively associated with economic growth [60]. However, research also indicates that the rate of technological innovation is contingent upon national economic development [61,62]. In times of economic prosperity, economies tend to allocate more capital to technological innovation activities, whereas during economic downturns, capital tends to be redirected towards investment to stimulate economic growth. Therefore, a causal endogenous relationship exists between Fintech and economic resilience. Economic resilience, as a macro-level variable, is influenced by multiple factors, and the presence of control variables in the current data may not prevent the occurrence of omitted variables. Potential issues of reverse causality and omitted variable bias necessitate an instrumental variable approach to addressing endogeneity and isolating the true effect of Fintech on economic resilience.

With reference to Ding et al. [12], the instrumental variable chosen to measure the level of Fintech development in this study is the geographical distance from each prefecture-level city to Hangzhou. This is because Hangzhou is the headquarters of Alibaba, the largest Fintech service provider in China with a highly developed Fintech level. By using geographic distance, which is a physical variable, the authors argue that this variable is less correlated with economic resilience and consistent with the characteristics of instrumental variable selection. To further strengthen the validity of the instrumental variable selection, we logarithmically processed the instrumental variables to eliminate the influence of magnitude. The results of the instrumental variables are presented in Table 4, and the test results are consistent with the benchmark regression. The use of geographic distance as an instrumental variable is particularly noteworthy, as it is less likely to be affected by confounding factors that may affect the relationship between Fintech development and other economic variables.

Table 4. Endogeneity test.

	(1)	(2)
	Fintech	ER
ER	0.622 ***	
	(3.25)	
Hdis		−0.112 ***
		(−3.23)
FD	−0.084	0.333 ***
	(−1.29)	(16.36)
Open	0.383 ***	−0.148
	(3.34)	(−0.80)
HC	−0.173 ***	0.090 ***
	(−9.42)	(8.92)
FS	1.062 *	−2.642 ***
	(1.86)	(−10.69)
_cons	5.453 ***	0.247
	(53.83)	(1.06)
N	2573	2573
R^2	0.384	0.161

Note: Robust t-statistics in brackets. *** $p < 0.01$, * $p < 0.10$.

5.4. Robustness Test

5.4.1. Alternative Proxies

The Fintech index can be classified into three dimensions: coverage breadth, usage depth, and digitization level [59]. Coverage breadth refers to the extent of electronic account coverage, including major Internet payment accounts and the number of associated bank accounts. Usage depth pertains to the actual use of Internet financial services in various areas, such as credit, investment, insurance, and payment. Digitization level, on the other hand, represents the mobility, affordability, convenience, and credit characteristics of Fintech innovation. To assess the impact of Fintech on economic resilience, we conducted regression analyses between each of the three Fintech dimensions and economic resilience, as presented in Table 5. The findings indicate that Fintech continues to play a significant positive role in bolstering economic resilience.

Table 5. Alternative proxies.

	(1)	(2)	(3)
	ER	ER	ER
Breadth	0.669 ***		
	(16.91)		
Depth		0.699 ***	
		(13.05)	
Digitization			0.584 ***
			(13.79)
FD	0.221 ***	0.263 ***	0.292 ***
	(7.21)	(8.90)	(9.70)
Open	−0.298	−0.343	−0.246
	(−1.02)	(−1.18)	(−0.86)
HC	0.168 ***	0.179 ***	0.167 ***
	(13.94)	(13.86)	(13.81)
FS	−2.074 ***	−2.271 ***	−2.987 ***
	(−8.47)	(−9.15)	(−12.36)
_cons	−3.862 ***	−4.123 ***	−3.563 ***
	(−18.86)	(−14.53)	(−15.61)
N	2573	2573	2573
R^2	0.219	0.201	0.217

Note: Robust t-statistics in brackets. *** $p < 0.01$.

5.4.2. Fixed Effect

In addition, to verify the robustness of our results, we also use a fixed model for regression, adding fixed effects to control for multiple unobservable local specific features to better verify the accuracy of our results. It should be noted that in the fixed effects test, we only control for city fixed effects but not time fixed effects because our economic resilience is calculated based on temporal variation, and controlling for time fixed effects would compromise the validity of economic resilience. The results are shown in Table 6. In Column (1), we control for city fixed effects only; in Column (2), we control for city-related factors and control for city fixed effects. The results in both columns remain consistent with the baseline regression.

Table 6. Individual fixed effects test.

	(1)	(2)
	ER	ER
Fintech	0.190 ***	0.761 ***
	(9.91)	(19.74)
FD		−0.028
		(−0.44)
Open		−1.641 ***
		(−8.07)
HC		0.139 ***
		(17.79)
FS		1.615 **
		(1.99)
_cons	−0.956 ***	−4.370 ***
	(−9.92)	(−20.08)
N	2579	2579
R^2	0.317	0.080

Note: Robust t-statistics in brackets. *** $p < 0.01$, ** $p < 0.05$.

5.5. Mechanism Test

Drawing on theoretical insights, we firstly investigate technological innovation as a mediating mechanism. The efficacy of science and technology innovation is fundamentally determined by the actual benefits it produces [63]. Patents, as tangible outcomes of innovation, are highly indicative of the level of regional science and technology innovation. Furthermore, the utilization of patents and the technology spillover effects that stem from them can be readily applied to the market and translated into tangible outcomes. Consequently, we follow Chen et al.'s [64] research and utilize the natural logarithm of the number of granted patents (R&D) in prefecture-level cities as a proxy variable for science and technology innovation. The larger the number of granted patents, the higher the level of science and technology innovation.

Secondly, we examine industrial structure as another mediating mechanism. Given the technological characteristics of Fintech, we posit that the optimization of the industrial structure driven by Fintech development should manifest in the high-end of the industrial structure, i.e., the proportion of strategic emerging industries should continue to increase, and high value-added industries should continue to emerge. The development of digital technology propels the economic structure towards the service industry; thus, the process of upgrading the high-end industrial structure is essentially the process of increasing the output value of the tertiary industry. Consequently, we adopt the proportion of the value added of the tertiary industry to the value added of the secondary industry to measure industrial structure upgrading (IS). Table 7 presents the findings of the mechanism test, which suggests that Fintech has a positive impact on technological innovation and industrial structure upgrading.

Table 7. Mechanism test.

	(1)	(2)
	R&D	IS
Fintech	1.656 ***	0.134 ***
	(23.95)	(6.69)
FD	0.095 *	0.238 ***
	(1.90)	(15.76)
Open	9.571 ***	0.146
	(15.47)	(1.54)
HC	−0.669 ***	−0.026 ***
	(−27.33)	(−5.46)
FS	−1.501 ***	1.404 ***
	(−2.94)	(10.79)
_cons	−4.410 ***	−0.460 ***
	(−12.32)	(−4.31)
N	2573	2573
R^2	0.667	0.459

Note: Robust t-statistics in brackets. *** $p < 0.01$, * $p < 0.10$.

5.6. Heterogeneity Test

China's regional economy displays a marked degree of spatial heterogeneity, with the eastern region exhibiting robust development, while the middle and western regions manifest weaker growth. Concurrently, the growth of Fintech in China is rapid, yet significant disparities exist among prefectures and cities, yielding an imbalanced spatial distribution featuring an evident step-like pattern at the regional level. These regional distinctions may also engender regional-level variations in the influence of Fintech on economic resilience. To further scrutinize the impact of Fintech on economic resilience at the regional level, this study focuses on two facets, namely the region and city scales.

To begin with, we adopt a regional/locational perspective by categorizing Chinese prefecture-level cities into eastern, central, western, and northeastern regions (Figure 1) and analyzing them through grouped regressions. Our findings, as presented in Table 8, reveal that Fintech exerts a notable impact on economic resilience, with a more significant effect observed in the east in comparison to the central and western regions. This differential impact may be attributed to the eastern region's inherent location advantages and resource endowments, as well as its sound economic development foundation and robust innovation activities, all of which enhance the promotional effects of Fintech on economic resilience. In contrast, Column (4) demonstrates the negative impact of Fintech on economic resilience in the northeastern region, which has traditionally served as a prominent industrial base in China, with a comparatively slow development of the financial sector. Consequently, the impact of Fintech on economic resilience in the region is not substantial.

Table 8. Locational heterogeneity.

	(1)	(2)	(3)	(4)
	East	Central	West	Northeast
Fintech	1.011 ***	0.987 ***	0.910 ***	−0.173 *
	(8.80)	(14.31)	(11.58)	(−1.69)
FD	0.529 ***	0.207 ***	0.038	0.178 ***
	(7.25)	(4.11)	(1.06)	(3.56)
Open	−1.094 ***	−5.162 ***	−0.214	0.469
	(−2.71)	(−4.62)	(−0.47)	(0.86)
HC	0.165 ***	0.165 ***	0.199 ***	0.074 ***
	(6.49)	(7.77)	(10.08)	(3.15)
FS	−5.191 ***	−3.773 ***	−1.334 ***	0.574
	(−6.24)	(−5.38)	(−4.24)	(1.23)
_cons	−5.648 ***	−5.157 ***	−5.012 ***	−0.092
	(−10.13)	(−14.20)	(−12.23)	(−0.19)
N	777	720	776	306
R^2	0.313	0.268	0.189	0.068

Note: Robust t-statistics in brackets. *** $p < 0.01$, * $p < 0.10$.

Table 9 present the results of the impact of Fintech on economic resilience in cities of different sizes. The findings indicate a positive and significant effect of Fintech on resilience across all city size classes, while the impact is most significant in large cities, followed by medium cities, and finally small cities. One possible explanation for this pattern is that larger cities tend to have greater economies of scale, which can promote technological innovation and foster the development of Fintech. In turn, the greater adoption of Fintech in larger cities may enhance economic resilience by improving access to financial services, facilitating entrepreneurship and innovation, and boosting overall economic growth. The heterogeneity results in this study are consistent with other relevant empirical studies [65].

Table 9. City-scale heterogeneity.

	(1)	(2)	(3)
	Small	Median	Big
Fintech	0.667 ***	0.711 ***	1.177 ***
	(8.41)	(11.80)	(11.45)
FD	−0.026	0.022	0.224 ***
	(−0.39)	(0.58)	(5.07)
Open	−1.145 **	−1.585 ***	0.131
	(−2.51)	(−4.66)	(0.29)
HC	0.157 ***	0.133 ***	0.219 ***
	(9.42)	(9.46)	(8.71)
FS	−0.151	−1.363 ***	−2.937 ***
	(−0.41)	(−2.86)	(−3.49)
_cons	−3.934 ***	−3.827 ***	−6.354 ***
	(−10.21)	(−12.53)	(−11.30)
N	715	928	936
R^2	0.120	0.143	0.224

Note: Robust t-statistics in brackets. *** $p < 0.01$, ** $p < 0.05$.

6. Conclusions and Enlightenment

This study ventured to enhance urban sustainability by examining economic resilience and digital transformation, aiming to broaden and redefine the traditional perception of economic resilience beyond mere shock response. It delved into economic resilience within the digital transformation of the financial technology sector, proposing a fresh perspective on navigating the complexities of modern urban development. We first introduced and employed an innovative conceptualization of economic resilience, framing it as an inherent attribute of regional economic systems. This novel perspective facilitates the analysis of economic resilience not only in the context of external disturbances but also as a fundamental characteristic, enabling a proactive assessment of a system's inherent economic stability and robustness. This shift towards recognizing resilience as a constant feature provides deeper insights into the foundational strength of economic systems, allowing for a more comprehensive understanding of their capacity to withstand and adapt to changes.

Secondly, this study developed a new method for quantifying economic resilience that adeptly captures the sensitivity and adaptability of economic systems without the prerequisite of a shock. This methodological innovation represents a significant leap forward in the quantification of economic resilience, offering a tool that can be applied universally, across various contexts and scenarios. Our empirical investigation, utilizing panel data from 286 prefecture-level cities in China from 2011 to 2020, demonstrates the utility of this method in examining the influence of Fintech on economic resilience and sheds light on the underlying mechanisms and the regional heterogeneity of this influence.

The empirical findings of this study reveal that the digital tools of Fintech significantly bolster economic resilience, serving as a new source of resilience through their facilitation of technological innovation and industrial structure upgrading. This underscores the critical role of Fintech in supporting the recovery, restructuring, and renewal of economies, thereby promoting high-quality economic growth. Our research aligns with and extends the findings of Zhou et al. [58] and Shi et al. [66], reinforcing the idea that Fintech is a pivotal catalyst for sustainable economic development.

However, this study is not without its limitations. Primarily, the measurement of economic resilience in this analysis was anchored solely on GDP as the input indicator. This choice, albeit grounded in the widespread availability and comparability of GDP data, may not fully encapsulate the multifaceted nature of economic resilience. Future studies could enrich our understanding of economic resilience by incorporating a broader spectrum of economic indicators, such as employment rates, industrial diversity, and innovation metrics. These additional indicators could offer a more nuanced and comprehensive view of the economic system's resilience, capturing aspects of economic health and adaptability beyond mere output.

Furthermore, the methodology employed to calculate economic resilience indicators relied on annual aggregated GDP data, leading to a uniform adaptability value being assigned to each city throughout the study period. This approach, while facilitating a streamlined analysis, may not accurately reflect the dynamic nature of cities' economic resilience, which can fluctuate significantly within shorter time frames due to various factors, including policy changes, market shifts, and external shocks. To address this limitation and enhance the granularity of resilience assessment, future research could leverage monthly data, or other more frequent economic indicators, to better capture the temporal variations in city's adaptability values. Such an approach would allow for a more detailed and responsive analysis of economic resilience, providing insights into the immediate impacts of economic policies and external events on regional economies.

In summary, this research makes significant strides in the conceptualization and quantification of economic resilience and further applies this methodology to empirical examination, highlighting the transformative potential of Fintech in enhancing the resilience of economic systems. While acknowledging the limitations of the current study, we advocate for further exploration into this complex relationship, suggesting that the economic resilience calculation method proposed here can be adapted and applied in diverse contexts to explore various dimensions of resilience. This study not only contributes to academic discourse but also offers practical insights for policymakers and practitioners interested in leveraging digital transitions to fortify urban sustainability across different scales and regions.

Author Contributions: Conceptualization and writing—review and editing, C.S.; formal analysis and writing—original draft, J.L. All authors have read and agreed to the published version of the manuscript.

Funding: This research was funded by the National Natural Science Foundation of China, grant number 72104149; R&D Program of Beijing Municipal Education Commission, grant number SM202310038007; the Academy of Metropolis Economic and Social Development at the Capital University of Economics and Business, grant number ZSXM2021003.

Data Availability Statement: The data supporting the conclusions of this article are available upon request.

Conflicts of Interest: The authors declare no conflicts of interest. The funders had no role in the design of the study; in the collection, analyses, or interpretation of the data; in the writing of the manuscript; or in the decision to publish the results.

Notes

1 There are various ways to define Fintech; according to the World Bank, Fintech is broadly defined as "advances in technology that have the potential to transform the provision of financial services, spurring the development of new business models, applications, and processes, and products" [11].

2 For different types of resilience analysis methodologies, refer to Doran and Fingleton [36] and Modica and Reggiani [37].

3 The exclusion of 2017 was due to the unavailability of GDP data from the China City Statistical Yearbook.

4 To eliminate the influence of the data outline, we add "1" to the original index and take the natural logarithm as the proxy variable of Fintech in this study. The larger the value, the higher the level of Fintech.

References

1. Bristow, G.; Healy, A. Introduction to the Handbook on Regional Economic Resilience. In *Handbook on Regional Economic Resilience*; Edward Elgar Publishing: Cheltenham, UK, 2020.
2. Ezcurra, R.; Rios, V. Quality of Government and Regional Resilience in the European Union. Evidence from the Great Recession. *Pap. Reg. Sci.* **2019**, *98*, 1267–1290. [CrossRef]
3. Martin, R.; Sunley, P.; Gardiner, B.; Tyler, P. How Regions React to Recessions: Resilience and the Role of Economic Structure. *Reg. Stud.* **2016**, *50*, 561–585. [CrossRef]
4. Holling, C.S. Resilience and Stability of Ecological Systems. *Annu. Rev. Ecol. Syst.* **1973**, *4*, 1–23. [CrossRef]
5. Martin, R.; Sunley, P. On the Notion of Regional Economic Resilience: Conceptualization and Explanation. *J. Econ. Geogr.* **2015**, *15*, 1–42. [CrossRef]
6. Wang, Z.; Wei, W. Regional Economic Resilience in China: Measurement and Determinants. *Reg. Stud.* **2021**, *55*, 1228–1239. [CrossRef]
7. Philibert Petit, E. Chapter 27—Smart City Technologies plus Nature-Based Solutions: Viable and Valuable Resources for Urban Resilience. In *Smart Cities Policies and Financing*; Vacca, R.J., Ed.; Elsevier: Amsterdam, The Netherlands, 2022; pp. 377–398. ISBN 978-0-12-819130-9.
8. Zhang, X.; Li, H. Urban Resilience and Urban Sustainability: What We Know and What Do Not Know? *Cities* **2018**, *72*, 141–148. [CrossRef]
9. Christopherson, S.; Michie, J.; Tyler, P. Regional Resilience: Theoretical and Empirical Perspectives. *Camb. J. Reg. Econ. Soc.* **2010**, *3*, 3–10. [CrossRef]
10. Tan, J.; Hu, X.; Hassink, R.; Ni, J. Industrial Structure or Agency: What Affects Regional Economic Resilience? Evidence from Resource-Based Cities in China. *Cities* **2020**, *106*, 102906. [CrossRef]
11. Feyen, E.; Natarajan, N.; Saal, M. *Fintech and the Future of Finance: Market and Policy Implications*; World Bank Group: Washington, DC, USA, 2023.
12. Ding, N.; Gu, L.; Peng, Y. Fintech, Financial Constraints and Innovation: Evidence from China. *J. Corp. Financ.* **2022**, *73*, 102194. [CrossRef]
13. Gomber, P.; Kauffman, R.J.; Parker, C.; Weber, B.W. On the Fintech Revolution: Interpreting the Forces of Innovation, Disruption, and Transformation in Financial Services. *J. Manag. Inf. Syst.* **2018**, *35*, 220–265. [CrossRef]
14. Alt, R.; Beck, R.; Smits, M.T. FinTech and the Transformation of the Financial Industry. *Electron Mark.* **2018**, *28*, 235–243. [CrossRef]

15. World Bank Fintech and the Future of Finance Overview Paper. 2022. Available online: https://www.worldbank.org/en/publication/fintech-and-the-future-of-finance (accessed on 5 April 2024).
16. Nguyen, C.P.; Su, T.D. Easing Economic Vulnerability: Multidimensional Evidence of Financial Development. *Q. Rev. Econ. Financ.* **2021**, *81*, 237–252. [CrossRef]
17. Rocchetta, S.; Mina, A.; Lee, C.; Kogler, D.F. Technological Knowledge Spaces and the Resilience of European Regions. *J. Econ. Geogr.* **2022**, *22*, 27–51. [CrossRef]
18. Tóth, G.; Elekes, Z.; Whittle, A.; Lee, C.; Kogler, D.F. Technology Network Structure Conditions the Economic Resilience of Regions. *Econ. Geogr.* **2022**, *98*, 355–378. [CrossRef]
19. Zhou, Q.; Zhu, M.; Qiao, Y.; Zhang, X.; Chen, J. Achieving Resilience through Smart Cities? Evidence from China. *Habitat Int.* **2021**, *111*, 102348. [CrossRef]
20. Yu, Z.; Li, Y.; Dai, L. Digital Finance and Regional Economic Resilience: Theoretical Framework and Empirical Test. *Financ. Res. Lett.* **2023**, *55*, 103920. [CrossRef]
21. Du, Y.; Wang, Q.; Zhou, J. How Does Digital Inclusive Finance Affect Economic Resilience: Evidence from 285 Cities in China. *Int. Rev. Financ. Anal.* **2023**, *88*, 102709. [CrossRef]
22. Hassink, R. Regional Resilience: A Promising Concept to Explain Differences in Regional Economic Adaptability? *Camb. J. Reg. Econ. Soc.* **2010**, *3*, 45–58. [CrossRef]
23. Martin, R. Regional Economic Resilience, Hysteresis and Recessionary Shocks. *J. Econ. Geogr.* **2012**, *12*, 1–32. [CrossRef]
24. Boschma, R. Towards an Evolutionary Perspective on Regional Resilience. *Reg. Stud.* **2015**, *49*, 733–751. [CrossRef]
25. Bristow, G.; Healy, A. Innovation and Regional Economic Resilience: An Exploratory Analysis. *Ann. Reg. Sci.* **2018**, *60*, 265–284. [CrossRef]
26. Hu, X.; Hassink, R. Adaptation, Adaptability and Regional Economic Resilience: A Conceptual Framework. In *Handbook on Regional Economic Resilience*; Bristow, G., Healy, A., Eds.; Edward Elgar Publishing: Cheltenham, UK, 2020; ISBN 978-1-78536-086-2.
27. Evenhuis, E. New Directions in Researching Regional Economic Resilience and Adaptation. *Geogr. Compass* **2017**, *11*, e12333. [CrossRef]
28. Martin, R.; Sunley, P. Regional Economic Resilience: Evolution and Evaluation. In *Handbook on Regional Economic Resilience*; Bristow, G., Healy, A., Eds.; Edward Elgar Publishing: Cheltenham, UK, 2020; ISBN 978-1-78536-086-2.
29. Sutton, J.; Arcidiacono, A.; Torrisi, G.; Arku, R.N. Regional Economic Resilience: A Scoping Review. *Prog. Hum. Geogr.* **2023**, *47*, 500–532. [CrossRef]
30. Sutton, J.; Arku, G. Regional Economic Resilience: Towards a System Approach. *Reg. Stud. Reg. Sci.* **2022**, *9*, 497–512. [CrossRef]
31. Pendall, R.; Foster, K.A.; Cowell, M. Resilience and Regions: Building Understanding of the Metaphor. *Camb. J. Reg. Econ. Soc.* **2010**, *3*, 71–84. [CrossRef]
32. Capello, R.; Caragliu, A.; Fratesi, U. Spatial Heterogeneity in the Costs of the Economic Crisis in Europe: Are Cities Sources of Regional Resilience? *J. Econ. Geogr.* **2015**, *15*, 951–972. [CrossRef]
33. Wang, S.; Scorsone, E. Economic Resilience after the Great Recession: An Examination of Unreserved Fund Balance in Michigan Counties. *Local Gov. Stud.* **2020**, *46*, 716–733. [CrossRef]
34. Yamamoto, D. Regional Resilience: Prospects for Regional Development Research: Regional Resilience. *Geogr. Compass* **2011**, *5*, 723–736. [CrossRef]
35. Simmie, J.; Martin, R. The Economic Resilience of Regions: Towards an Evolutionary Approach. *Camb. J. Reg. Econ. Soc.* **2010**, *3*, 27–43. [CrossRef]
36. Doran, J.; Fingleton, B. Employment Resilience in Europe and the 2008 Economic Crisis: Insights from Micro-Level Data. *Reg. Stud.* **2016**, *50*, 644–656. [CrossRef]
37. Modica, M.; Reggiani, A. Spatial Economic Resilience: Overview and Perspectives. *Netw. Spat. Econ.* **2015**, *15*, 211–233. [CrossRef]
38. Hu, X.; Li, L.; Dong, K. What Matters for Regional Economic Resilience amid COVID-19? Evidence from Cities in Northeast China. *Cities* **2022**, *120*, 103440. [CrossRef] [PubMed]
39. Mai, X.; Zhan, C.; Chan, R.C.K. The Nexus between (Re)Production of Space and Economic Resilience: An Analysis of Chinese Cities. *Habitat Int.* **2021**, *109*, 102326. [CrossRef]
40. Kusumastuti, R.D.; Husodo, Z.A.; Suardi, L.; Danarsari, D.N. Developing a Resilience Index towards Natural Disasters in Indonesia. *Int. J. Disaster Risk Reduct.* **2014**, *10*, 327–340. [CrossRef]
41. Rose, A.; Krausmann, E. An Economic Framework for the Development of a Resilience Index for Business Recovery. *Int. J. Disaster Risk Reduct.* **2013**, *5*, 73–83. [CrossRef]
42. Vella, L.B. Gordon Cordina, Nadia Farrugia, Stephanie Economic Vulnerability and Resilience: Concepts and Measurements. In *Measuring Vulnerability in Developing Countries*; Routledge: London, UK, 2012; ISBN 978-1-315-87437-1.
43. Shen, Y.; Hueng, C.J.; Hu, W. Using Digital Technology to Improve Financial Inclusion in China. *Appl. Econ. Lett.* **2020**, *27*, 30–34. [CrossRef]
44. Erel, I.; Liebersohn, J. Can FinTech Reduce Disparities in Access to Finance? Evidence from the Paycheck Protection Program. *J. Financ. Econ.* **2022**, *146*, 90–118. [CrossRef]
45. Feng, S.; Zhang, R.; Li, G. Environmental Decentralization, Digital Finance and Green Technology Innovation. *Struct. Change Econ. Dyn.* **2022**, *61*, 70–83. [CrossRef]
46. Anagnostopoulos, I. Fintech and Regtech: Impact on Regulators and Banks. *J. Econ. Bus.* **2018**, *100*, 7–25. [CrossRef]

47. Liu, Y.; Luan, L.; Wu, W.; Zhang, Z.; Hsu, Y. Can Digital Financial Inclusion Promote China's Economic Growth? *Int. Rev. Financ. Anal.* **2021**, *78*, 101889. [CrossRef]
48. Wang, Y.; Xiuping, S.; Zhang, Q. Can Fintech Improve the Efficiency of Commercial Banks? —An Analysis Based on Big Data. *Res. Int. Bus. Financ.* **2021**, *55*, 101338. [CrossRef]
49. Yang, L.; Wang, S. Do Fintech Applications Promote Regional Innovation Efficiency? Empirical Evidence from China. *Socio-Econ. Plan. Sci.* **2022**, *83*, 101258. [CrossRef]
50. Xue, Q.; Bai, C.; Xiao, W. Fintech and Corporate Green Technology Innovation: Impacts and Mechanisms. *Manag. Decis. Econ.* **2022**, *43*, 3898–3914. [CrossRef]
51. Mudambi, R.; Mudambi, S.M.; Mukherjee, D.; Scalera, V.G. Global Connectivity and the Evolution of Industrial Clusters: From Tires to Polymers in Northeast Ohio. *Ind. Mark. Manag.* **2017**, *61*, 20–29. [CrossRef]
52. Liao, J.; Kickul, J.R.; Ma, H. Organizational Dynamic Capability and Innovation: An Empirical Examination of Internet Firms. *J. Small Bus. Manag.* **2009**, *47*, 263–286. [CrossRef]
53. Ren, X.; Zeng, G.; Gozgor, G. How Does Digital Finance Affect Industrial Structure Upgrading? Evidence from Chinese Prefecture-Level Cities. *J. Environ. Manag.* **2023**, *330*, 117125. [CrossRef] [PubMed]
54. Ndubuisi, G.; Otioma, C.; Tetteh, G.K. Digital Infrastructure and Employment in Services: Evidence from Sub-Saharan African Countries. *Telecommun. Policy* **2021**, *45*, 102153. [CrossRef]
55. Fingleton, B.; Garretsen, H.; Martin, R. Recessionary Shocks and Regional Employment: Evidence on the Resilience of UK Regions*. *J. Reg. Sci.* **2012**, *52*, 109–133. [CrossRef]
56. Hu, X.; Hassink, R. Exploring Adaptation and Adaptability in Uneven Economic Resilience: A Tale of Two Chinese Mining Regions. *Camb. J. Reg. Econ. Soc.* **2017**, *10*, 527–541. [CrossRef]
57. Shi, C.; Guo, N.; Gao, X.; Wu, F. How Carbon Emission Reduction Is Going to Affect Urban Resilience. *J. Clean. Prod.* **2022**, *372*, 133737. [CrossRef]
58. Zhou, G.; Zhu, J.; Luo, S. The Impact of Fintech Innovation on Green Growth in China: Mediating Effect of Green Finance. *Ecol. Econ.* **2022**, *193*, 107308. [CrossRef]
59. Guo, F.; Wang, J.; Wang, F.; Kong, T.; Zhang, X.; Cheng, Z. Measuring China's Digital Financial Inclusion: Index Compilation and Spatial Characteristics (in Chinese). *China Econ. Q.* **2020**, *19*, 1401–1418.
60. Hu, Y.; Liu, C.; Peng, J. Financial Inclusion and Agricultural Total Factor Productivity Growth in China. *Econ. Model.* **2021**, *96*, 68–82. [CrossRef]
61. Xiao, S.; Zhao, S. Financial Development, Government Ownership of Banks and Firm Innovation. *J. Int. Money Financ.* **2012**, *31*, 880–906. [CrossRef]
62. Zhu, X.; Asimakopoulos, S.; Kim, J. Financial Development and Innovation-Led Growth: Is Too Much Finance Better? *J. Int. Money Financ.* **2020**, *100*, 102083. [CrossRef]
63. Rauter, R.; Globocnik, D.; Perl-Vorbach, E.; Baumgartner, R.J. Open Innovation and Its Effects on Economic and Sustainability Innovation Performance. *J. Innov. Knowl.* **2019**, *4*, 226–233. [CrossRef]
64. Chen, W.; Si, W.; Chen, Z.-M. How Technological Innovations Affect Urban Eco-Efficiency in China: A Prefecture-Level Panel Data Analysis. *J. Clean. Prod.* **2020**, *270*, 122479. [CrossRef]
65. Tian, Y.; Guo, L. Digital Development and the Improvement of Urban Economic Resilience: Evidence from China. *Heliyon* **2023**, *9*, e21087. [CrossRef]
66. Shi, Y.; Zhang, T.; Jiang, Y. Digital Economy, Technological Innovation and Urban Resilience. *Sustainability* **2023**, *15*, 9250. [CrossRef]

land

MDPI

Article

Enabling In-Situ Urbanization through Digitalization

Le Li [1] and Tao Song [2,3,*]

[1] Rural Development Institute, Chinese Acadmy of Social Sciences, Beijing 100732, China; nfs-lile@cass.org.cn
[2] Institute of Geographic Sciences and Natural Resources Research, Chinese Academy of Sciences, Beijing 100101, China
[3] University of Chinese Academy of Sciences, Beijing 100049, China
* Correspondence: songtao@igsnrr.ac.cn

Abstract: The bourgeoning of e-commerce in the context of the information era has accelerated the urbanization trend by broaching a new horizon of economic and industrial boosters for rural places, epitomized by a great number of "Taobao Villages" in China. This paper has two objectives: (1) explore the process and mechanism of digitalization enabling rural in-situ urbanization represented by e-commerce; (2) nuance the specific case evidence of Daiji Town, where digitalization enabled in-situ urbanization recently. We build up a theoretical framework for digitalization-enabled in-situ urbanization from the juxtaposition of four interlinked elements: industry, talent, rural governance, and land use. It then analyzed the details and evidence of digitalization enabling rural in-situ urbanization through the case study of Daiji Town. The main conclusions of this paper are as follows: First, digitalization plugs rural areas into production and consumption networks in wider contexts, promoting the transformation and prosperity of rural economies. Secondly, the reverse migration of young generations to rural areas becomes the key to rural in-situ urbanization. Thirdly, digitization materializes the urbanization of rural spaces. Finally, digitalization enables the rural transformation and improvement of urban-rural relations in the Global South, which needs to be further explored.

Keywords: digitalization; rural in-situ urbanization; Taobao village; China

1. Introduction

The continuous prevalence of e-commerce in the UK, China, Africa, the BRICS countries, and other developed and developing countries is of great significance for the revitalization of rural areas in the world [1–3]. In the past decade, the digitalization represented by e-commerce has been reshaping China's rural society at an unimaginable speed [4]. Our attention to Cao County in Shandong Province stems from a news article vignette in May 2021, in which an online blogger believes that his hometown, Cao County, is more attractive than such modern metropolises as Beijing and Shanghai, attracting more than 260 million video views and over 500 million topics read on the microblog platform about Cao County [5]. The flow dividend has attracted widespread social attention to this little-known small county in the southwestern Shandong Province, China. There are nearly 60,000 online shops in the whole county, driving 350,000 people to start businesses and engage in relevant work. With 151 Taobao villages and 17 Taobao towns in the county and an annual sale of more than 20 billion yuan, it becomes the second largest rural e-commerce cluster in China after Yiwu in Zhejiang Province [6]. Taobao, owned by the electronic retailer Alibaba Group, is the largest e-commerce platform in China, along with eBay and Amazon. "Taobao Village" is a cluster of rural electronic retailers in an administrative village. With more than 10% of households operating online Taobao stores, the total turnover there exceeds 10 million yuan [7]. And when a town has three Taobao villages, or its annual e-commerce sales from them exceed 30 million yuan, and the number of active online shops exceeds 300, it could be termed a Taobao town [8]. In 2021, there will be a total of

7023 Taobao villages and 2171 Taobao towns in China, widely distributed in 27 provincial administrative units [9].

The wonders generated by the combination of e-commerce and China's rural and small towns, with Taobao Village and Taobao Town as carriers, have quickly attracted the interest of the academic community. The existing literature has studied the industrial and production structure [4], employment opportunities [10], community organization and governance [11], lifestyle [12], rural environment [13], and spatial structure [14–16] of Taobao Villages. A vast number of rural geographical scholars have noticed that digital transformation is radically changing the constellations and processes of production, marketing, and consumption in the rural production system [17]. On the one hand, digital information technology has changed the information asymmetry between supply and demand in rural areas and the outside world, making farmers' production and management targeted [18]. On the other hand, it promotes the links between rural industries by reorganizing the original production factors and rural communities [19]. E-commerce has become a technical catalyst for changes in rural industrial structure, employment patterns, and household economies [10]. Moreover, noticeable evidence proves that rurality and everyday mundane life in rural China have changed in response to skyrocketing digitalization [12], especially in burgeoning Taobao villages in China's coastal areas, which serves as a prime example of the global "digital turn". In essence, this change can be seen as the impact of digitalization on the spatial reconstruction of rural economic geography.

The process of drawing urban cores and even distant hinterlands into orbit, thus giving them more of the physical, economic, social, and cultural characteristics usually associated with cities, is one of the most commonly studied processes within scholarship on planetary urbanization in the global South [20–24]. How migration contributes to the urbanization process has attracted more attention among a vast array of theoretical urbanization inquiries [25–27]. Migrants have traditionally been viewed as being responsible for excessive urban growth and urban surplus labor [28–33]. However, since the 1980s, China has been witnessing the emergence of rapid rural industrialization in the form of township and village enterprisesenterprises (TVEs), which not only led to the rapid development of towns but also the huge expansion of villages. Scholars have termed this phenomenon "in situ urbanization" or "urbanization from below", distinguishing it from city-centered urbanization [34–37]. Thus, in-situ urbanization is a process by which peasants harvest non-agricultural employment and citizenization based on the original village and small towns [38] by engaging in rurally diversified economies [39]. This rural economic boom engenders sweeping improvements in physical infrastructure such as roads, electricity, and information technology, as well as transforming the countryside into a "functional extension of the city" [40]. Later on, as economic reforms shifted the focus from rural to urban areas, especially after the 1993 recentralization of public finances, TVEs and townships were phased out of the central stage of the policy agenda in China [41]. Beyond these vignettes of rural urbanization through industrialization, the stories of digitalized in-situ urbanization remain poorly discussed.

Up to date, under the guidance of China's national scheme towards rural revitalization, a new round of rural urbanization is fundamentally reconstructing rural settlements [42,43]. The above research showed that e-commerce, by using the time-space compression effect, had built a bridge between rural producers and urban consumers and accelerated the reallocation of capital, personnel, wealth, and information between urban and rural areas [44]. Rural production and lifestyle are "embedded" into urban life through the Internet, forming a benign interaction between rural and urban areas. This interaction is a complex and comprehensive process that reconstructs the rural economic structure, social form, governance mode, and spatial structure [45,46], showing the typical significance of rural urbanization.

It is noteworthy that there are already some discussions about the rural in-situ urbanization in Taobao Village. Luo and He believed that Taobao Village is the product of e-commerce in the information age, revealing a new type of rural in-situ urbanization [47]. This process reconstructs the social, economic, and physical spaces of rural areas. Lou and

Hu proposed the concept of "urbanization in real time", trying to explain that urbanization can be realized without completely changing the form of rural settlements through real-time communication technology and rapid transportation [48]. Lin constructed an analytical framework of e-urbanism, which consists of three interwoven levels: the ICT infrastructure and production network, the social network and power relationship, and the urban form and land use [49]. Although the above research has noted the signs of urbanization in Taobao Village, it still lacks empirical and qualitative evidence of the impacts of digitalization on urbanization. Much as digitalization has a central place in geographical inquiries, a holistic and nuanced analysis of digitalization-enabled rural in-situ urbanization coupled with erasing the binary between urbanization and rurality is urgently needed.

Digitalization is looming large as a leviathan booster of global rural restructuring by enabling urbanization. Nonetheless, there is still a lack of a framework based on a comprehensive perspective to describe and depict the complex and overall process of the role mechanism of digitalization enabling in-situ urbanization, especially in the context of China's national scheme towards rural revitalization. This paper has two objectives: (1) explore the process and mechanism of digitalization enabling in-situ urbanization represented by e-commerce and try to verify them through microscopic empirical research. (2) to nuance the specific case evidence of Daiji Town, where digitalization-enabled urbanization is starting recently. It is structured as follows: After the introduction, we will propose a theoretical framework for digitalization-enabled urbanization in the second part. The third part will introduce methods and data collection. The profile of our case evidence will be shown in the fourth part. It then elaborates on the nuanced process and mechanism of digitalization-enabled urbanization in Daiji Town. We conclude with some key highlights and policy discussions.

2. Digitalization-Enabled Urbanization in Rural China towards a Conceptual Framework

In this subsection, we curate the digitalization-enabled urbanization framework from the juxtaposition of four interlinked elements (Figure 1), which echo recent scholarly buzzwords and concomitantly construct the digitalization-enabled urbanization theoretical landscape.

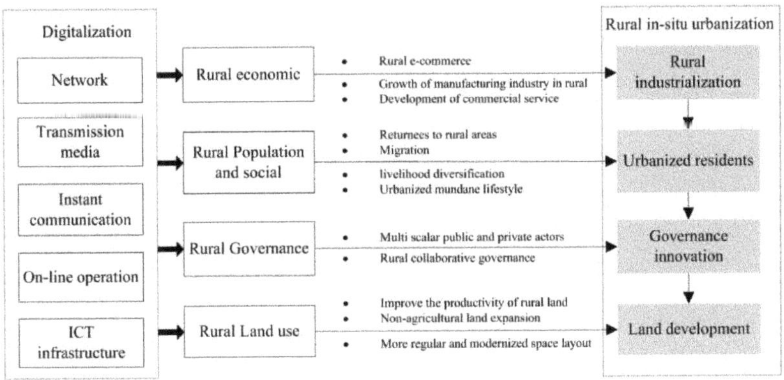

Figure 1. Digitalization-enabled rural in-situ urbanization.

2.1. *Digitalization-Enabled Industrialization*

Digitalization is regarded as broaching a new horizon of economic and industrial boosters for rural places. Evidence proves that the digital transformation is radically changing constellations and processes of production, marketing, and consumption in the agrifood system, helping farmers deliver safe, sustainable, and quality food [50]. However, it can also be more adaptive to climate change [51]. In addition to upgrading traditional agriculture, digitalization-enabled rural growth is more prominent and empowered by the so-called "rural industry", which refers to the dominant non-agricultural industry, either in manufacture or services, in one township and surrounding villages [13]. As China's evidence shows, family-based workshops facilitated early rural industrialization, which then turned into urban agglomerations [52]. To date, the overall development of rural industry in rural China has been rapid and spatially uneven, with highly developed industries in the eastern region and less developed industries in the middle and western regions [53]. The Taobao village, featured by diffusing e-commerce platforms in rural China, is regarded as a prominent and effective means of revitalizing rural areas and narrowing the rural-urban gap by both academia and the government [13], through transforming the structure, processes, and strategies of most industries but also spawning entirely new enterprises and industrial chains. In addition to China's case evidence, Germany's tales also proved that digital platform-oriented firms are relatively often located in urban regions, whereas digital manufacturers are of significance for rural industrialization [54].

2.2. *Digitalization-Enabled Urbanized Residents*

Instead of passive donation recipients and price takers, farmers serve as active entrepreneurs who take risks and innovate to adapt to the transition from farmers to urbanized residents [55]. In this vein, rural residents are enabled through diverse innovative practices and knowledge-intensive skills [56]. The diversification of the economic base of rural communities has often been read as a shift toward post-productivism [57,58]. Accordingly, farmers, immigrants, and second-home owners living in rural areas reap momentum equipped with digital skills and sufficient funding in this mixed post-productivism milieu and thus play a proactive agency role in the rural progression featured by the intertwined and mutually constitutive production of hybridity, diversity, and heterogeneity [59]. Indeed, the "silent revolution" of digitalization in rural China is transforming and reconstituting both rural economies and civil society. As Wang et al. illustrated, the multifaceted nature of rural transformation wreaked by e-commerce disintermediation includes livelihood diversification, the de-linking of livelihoods from land, the de-localization of livelihoods, cultural and social changes, and the disintermediation of e-commerce, all of which amount to complicating the deepening of the rural-urban divide. In sum, digitalization-enabled urbanization is a flexible relational process that is embedded in the mundane everyday practices of smallholder farmers, indigenous rural dwellers, and e-commerce employers and employees [60].

2.3. *Digitalization-Enabled Governance Innovation*

In the multiple digitalization-related definitions, it is common to refer to the application of digitalization to innovate governance through various technologies by multi-scalar actors. The technology industry's ascent as a force advancing urban governance ideology is particularly important because it permeates cities' political administrative structures and reframes public discourse about urban issues in technology-centered terms [61,62]. For example, the "Smart Territory" is termed the one that seeks to solve public problems through solutions based on technology in the framework of a partnership between multiple participants from different sectors, both public and private [63,64]. Accordingly, the digitalization dimension of a policy is not linked to technological determinism but instead refers to a broader context of abilities to ramp up governance innovation at some specific sites by both multi-scalar public and private actors. Moreover, governments at multiple scales around the world have diversified their policy portfolios to galvanize multiple economic, social, and environmental governance targets. Liang and Li noted that the Chinese government has issued a series of governance innovations and policies to promote the development of the digital economy [65]. Jamil unpacked digitization policy-related challenges in Pakistan, including a lack of policy evaluation and refinement, a lack of focused research, and inappropriate allocations of funds at federal, national, and sectoral levels that affect wide-ranging digitalization in Pakistan [66]. In this paper, we examine how digitalization has enabled governance agencies in rural areas through governance service delivery, community engagement, and multi-scalar civic participation.

2.4. Land Reconfigurations Driven by Digitalization

"Rural-urban digital divide" has been coined since the 21st century to picture and characterize spatial inequalities during the digital era [67]. Rosen and León developed the "digital growth machine" rationale to reveal the rural-urban land transition in pursuit of capturing land-related profit through traditional land expansions and digitalization intensification strategies [61]. Beyond simplistically replicating traditional land growth in a new digital sector, the digital growth machine regards the land as a panoply of physical and digitized manifestations that range from goods and services to data flows, whereby land use forms and intensification have been thoroughly reconfigured. To curb such spatial inequalities, which are of significance for rural land development, digital technologies and applications are widely discussed and aimed at improving the productivity of rural land, advancing rural economies, and favoring the inclusion of rural communities in cultural, social, and political activities. Digitalization through such digital technologies as sensors, cloud computing, smart grids, social media platforms, etc. has been actively engaging in the digitalized circularity of urban metabolism [68], cohesion policies, and sustainable land management [64].

Drawing on this conceptual assemblage, this approach delivers the opportunity to transcend the unitary dichotomy between rural/urban areas by integrating all potent elements into a holistic framework of digitalization-enabled urbanization. In tracing the subfield of this dynamic through the following case evidence, the paper places an emphasis on how various elements constitute and sustain the in-situ urbanization driven by digitalization.

3. Data and Method

Data supporting this research is collected and assembled in three distinct but interlinked ways: (1) first-hand data, which is mainly collected from in-depth interviews with local government officials, village cadres, online shop operators, general staff of e-commerce companies, and local villagers. (2) the second-hand data, including relevant government documents and plans, publications published in journals, newspapers, and the Internet, and media interviews and reports. (3) In addition to the spatial data, we also visually demonstrate the spatial layout of Taobao Village in Daiji Town by deploying the Arcgis model to discern land use changes caused by the development of the e-commerce industry there.

The authors of the manuscript conducted ethnographic fieldwork in Cao County and Daiji Town mainly between 2022 and 2023, with some short follow-up trips in 2023, in order to make an in-depth, conceptual, and theoretical description and analysis of the internal operation of rural socioeconomic structure and the process of digitalization that enabled in-situ urbanization. Semi-structured and open-ended interviews with more than 50 government officials, village cadres, enterprise managers, ordinary residents, factory workers, and villages were completed accordingly. Snowball sampling was conducted in these ways: first, through local connections by the first author, who lived in Daiji Town for about 2 months and worked in an embroidery studio for 3 weeks; and second, via official connections the other coauthor built up through the role of consultant for a project conducted in Shandong Province. Table 1 lists the key respondents' information in this study. From the interviews with government departments of Cao County and government officials of Daiji Town, we mainly learned information from three aspects: (1) the role of government departments in the development of digitalization and the e-commerce industry in Daiji Town; (2) the process of development, operation, and management in the E-commerce Industrial Zone of Daiji Town; and (3) the governments' planning and toolkit for the future development of e-commerce in Daiji Town. Moreover, we conducted interviews with relevant village cadres of Dinglou and Sunzhuang Village, which are the earliest to become Taobao villages in Daiji Town, trying to figure out the following three issues: (1) how e-commerce originated and rapidly developed in villages; (2) how the development of e-commerce changed and reshaped rural landscapes; and (3) how digitalization impacted rural governance. The study protocol was approved by the local government of Daiji Town. The study also recruited more than 20 village participants through face-to-face intercept interviews. All of these villagers were engaged in digitalized e-commerce businesses as bosses or employees. Several related questions were asked: (1) the basic situation and the production and management links of online shops; (2) the impacts of e-commerce and digitalization on their livelihood, income level, and living habits; and (3) their feelings about the surrounding environment and visions for the future of the village. 4. Profile of Daiji Town in Shandong Province.

Table 1. Key actors interviewed by the study.

Category	Name	Organization/Identity	Code
Government officials	Mr. Zhang	Cao County Government, E-commerce Service Center	GO1
	Mr. Xie	Cao County Government, Bureau of Industry and Information Technology	GO2
	Mr. Li	Daiji Town Government	GO3
	Mr. Li	Daiji Town Government	GO4
Village cadres	Mr. Sun	Sunzhuang Village	VC1
	Mr. Sun	Sunzhuang Village	VC2
	Mr. Reng	Dinglou Village	VC3
Entrepreneurs of leading enterprises	Mr. Hu	Executive Director, Chenfei Clothing Co., Ltd.	EL1
	Mrs. Meng	Design Director, Chen Fei Clothing Co., Ltd.	EL2
	Mrs. Zhou	Manager, Qingsheng Performance Clothing Co., Ltd.	EL3
Villager	Mrs. Cui	Owner of Hanfu online shop	V1
	Mr. Zhao	Owner of performance costumes online shop	V2
	Mr. Sun	Owner of performance costumes online shop	V3
	Mrs. Liu	Staff of fabric store	V4
	Mr. Zhang	Staff of Yunda Express	V5

Our case Daiji Town is located in the southeast of Cao County, Shandong Province, with a total area of 45 square kilometers and a total population of about 47,000 (Figure 2). Aquaculture, planting, and processing of agricultural and sideline products were the leading industries in the 1980s. In the late 1990s, Cao County actively opened up a number of textile factories and established vocational and technical schools to train textile skills and promote economic development; therefore, the villagers of Daiji mastered the skills of making performance costumes. In 2009, China's "online shopping" craze was in a period of rapid development, and the increase of cultural performance activities created a huge segment of the market for performance costumes. Daiji Town became one of the earliest Taobao Towns in China, and all 32 administrative villages under its jurisdiction have been identified as "Taobao Villages". According to data from the Ali Research Institute, a small town such as Daiji Town in Cao County, where every village has been rated as the "Taobao Village" for four consecutive years, is unique in the whole country [9].

As the development models of Taobao towns are so diverse that it seems impossible to use just one or two of them to fully summarize the more than 2000 Taobao towns in China, our case represents a typical example of digitalization-enabled urbanization in rural areas, which should be applicable to many small towns in China for the following reasons:

Firstly, e-commerce in Daiji Town is booming. Dominated by the performance costumes and Hanfu manufacture, the e-commerce market there is getting rapid growth, currently with 18,000 registered online shops, and about 70% of performance costumes and 30% of Hanfu in China's e-commerce market are produced in Daiji Town. In addition, e-commerce sales here have also shown explosive growth in recent years. According to interviews with the local government staff, the e-commerce transactions in Daiji Town reached

9 billion yuan in 2021, with about 80% of the local residents engaged in e-commerce-related jobs (GO1).

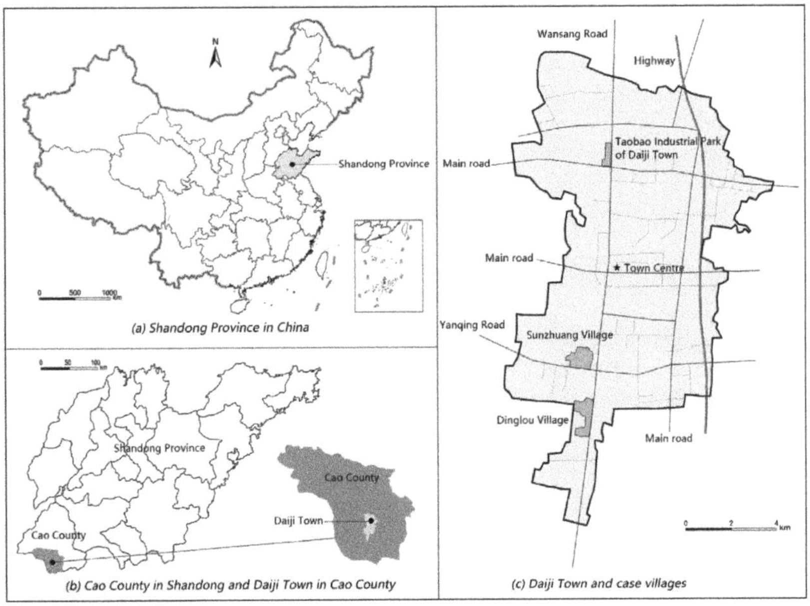

Figure 2. The location of the case study area.

Secondly, the initial development of e-commerce in Daiji Town is completely a result of spontaneous market rationality, which is dominated by local entrepreneurs to a large extent. And most Taobao villages and towns in China follow a similar bottom-up development path.

Thirdly, previous studies have shown that residents in villages close to urban centers can be more successful in diversifying their livelihoods by combining the non-agricultural employment opportunities nearby [69]. In our case, Daiji Town is a traditional agricultural town far away from the city center. With no national or provincial highways but only one county road passing through the border, this town does not have any advantage in its traffic conditions. And most of the villages in China have a similar development background and foundation. Therefore, our case study can help us understand how digitization empowers the urbanization of these agricultural towns far from the periphery of big cities.

4. Digitalization-Enabled Urbanization in Daiji Town

4.1. Digitalization, E-Commerce, and Industrialization

The traditional offline transaction is conducted through face-to-face communication: the buyers go to a specific place, communicate with the sellers, and complete the exchange of information, goods, and funds through direct contact. In contrast, e-commerce delivers a digital internet platform through which buyers browse the details of goods, communicate with sellers, and complete the payment. The sellers deliver the goods to the buyers through the logistics system shortly after confirming the payment. In short, e-commerce enables buyers and sellers to exchange information and trade over an infinitely wide geographical area, breaking the dependence of trade on spatial location and distance [70].

Our case evidence will demonstrate the nexus of digitalization, e-commerce, and industrialization by witnessing the vicissitudes of the performance costume market, which is a "niche market" with a small audience. Before the development of e-commerce, the sales of costumes in rural areas depended to a large extent on the market in the county, making the trade basically limited to the local area. As a result of the narrow sales channels, the costume manufacturing enterprises in Daiji Town were generally small in scale and low in output. Then the emergence of digitalized e-commerce broke the information asymmetry between small merchants and the outside market, allowing producers to directly have access to the super-sized potential market with the possibility of sufficient market demand to support all types of products. Through e-commerce, Daiji Town was connected to national markets, and buyers from different regions started to buy their products, making the sales increase significantly. Shortly afterward, thanks to the "ripple effect" in China's rural society [71], the information and experience of successful online trades spread rapidly among relatives and neighbors, and more and more rural households joined the online business.

"Twenty years ago, we usually took our costume samples to the wedding photography shops in the county town to sell them door-to-door. In case of product overstock, we never produced too much at one time. And in order to sell the costumes, the farthest place I went was Heilongjiang Province. But so hard as I tried, the number of costumes sold out in a year could be about 100 at most (EL3)".

"In 2009, one of my friends came to visit me and mentioned "Taobao", a platform where people could sell costumes without renting shops or going out for promotion. My wife decided to open an online shop there immediately. When the Children's Day on June 1st was coming, the first customer took the initiative to contact us online and ordered over 200 sets of student uniforms at one time. As well as excited, we were worried about having such a big order that we could not finish the costumes manufacturing in time, so we invited two neighbors to join us and in this way, we finished the first electronic order in Daiji Town (VC3)".

Digitalization creates a data-driven production mode, which changes the traditional way of production organization in rural areas. The development of social media and Web 2.0 provides great support for the transformation of e-commerce from product-oriented to social and customer-centric [4]. Digital communication has enabled designers, consumers, and small businesses to have direct contact opportunities with each other. The information and data fed back by the market became an important basis for the production of enterprises. Once market demand changes, the production cooperation between small and medium-sized enterprises and family workshops could be quickly adjusted and reorganized into a new production network. This model was beneficial for economic organizations with limited production capacity and marketing capacity in rural areas to avoid potential risks. What is more, the information exchange has also inspired designers and small enterprises to constantly create new styles and new categories of products to obtain new markets.

"I joined dozens of online communities established by the Hanfu enthusiasts, including WeChat group, QQ group and Douban group, to learn about the clothing style and price preference from young people. We would publish our design sketches in the communities to get real-time feedback from potential consumers, including the clothing styles, colors, fabrics and price. And the Hanfu enthusiasts could exchange any of their ideas with me here. Moreover, we would launch online polls and the most popular and potentially best-selling products would be massed-produced in our factory(V1)".

Whereas the structure, processes, and strategies of most industries have been reshaped by e-commerce, thousands of entirely new enterprises and industrial chains have sprung up in this trend. With the increasing number of online businesses in Daiji Town, many rural households began to set up enterprises of raw material and accessory businesses for performance costumes and Hanfu, enabling the clothing enterprises and family workshops to have local access to all the raw materials needed for production (Figure 3). The raw materials for these enterprises mainly came from Zhejiang Province, a major textile production area in China. At the same time, enterprises with some new functions also emerged. By providing professional webcasting services, online marketing strategies, and online or offline employee training for clothing manufacturing enterprises, they played key roles in accurately coordinating relationships between the upstream and the downstream, producers and consumers (Figure 4). The development of Taobao villages has enabled a large number of small, scattered businesses in rural areas of China to be connected and clustered into modern urban business assemblages through ICT and logistics networks [4].

(**a**) (**b**) (**c**)

Figure 3. Raw material and accessory businesses for performance costumes and Hanfu (Source: photos taken during the fieldwork). (**a**) raw material and accessory district, (**b**) Thread and tailor's shops, (**c**) Embroidering and textile printing shops.

Figure 4. A media company provides photo services for online shops (Source: photos taken during the fieldwork).

"On both sides of the Wansang Road, there are as many as 60 to 70 companies selling cloth and more than 200 companies running subsidiary material businesses. Moreover, here also distribute companies mainly engaged in designing, pattern making, embroidering, and textile printing. The Sunzhuang Village plans to expand the original scale and improve the upstream industrial chain of performance costume production. In the future, with the sales channels of performance costumes and Hanfu broadened, the demand and scope of the market for subsidiary materials will be larger, which urges us to speed up the pace of opening new cloth stores. Our goal is to become the largest subsidiary material market of costumes in the north of the Yangtze River (VC2)".

4.2. Reverse Migration and Social Changes Associated with Digitalization

The unprecedented worldwide digitization has not only unleashed economic miracles for a vast number of villages but also begat a series of profound changes to the social structure of these rural areas. In the context of the declining economic status of rural areas caused by the binary of urban-rural and the constant outflow of young and middle-aged population into urbanized areas, attracting talents back to the countryside has always been an important scheme of rural revitalization in China [72]. The booming development of digitalized e-commerce has spawned a larger number of jobs with considerable incomes.

As the neo-classical economic theory implies, higher incomes and better employment opportunities will drive population migration [4,73]. As a result, a reverse migration of the population began to take place, i.e., a large number of young people who had moved from rural areas to urban areas returned to the countryside [74].

According to our interview, Daiji Town used to be a big labor export town, with about 60% of the labor force migrating to cities to seek a better life. Since the rapid development of e-commerce in the 1990s, lots of new employment opportunities have been created, attracting the younger generation in rural areas back to their hometowns. Some of those jobs are directly related to e-commerce, such as those of online shopkeepers, website designers, Taobao photographers, and so on. while the others are indirectly related, including both the upstream and the downstream, such as logistics and distribution. According to local government officials, by the year 2021, there would have been more than 7000 migrant farmers and over 700 college graduates in Daiji Town coming back to start businesses, with 2 PhD students and 14 Master students included, which is very rare in rural areas.

"I used to work in clothing factories run by Koreans in Qingdao and Yantai with a salary of more than 3000 yuan a month. Now I and my wife operate a store on Taobao, which enables us to earn 150,000 yuan a year without too hard work. We can make no less money through business on Taobao than working outside and we can take better care of the elderly and children at home(V2)".

A large number of young people returning to the countryside have reinvigorated the languishing rural areas and boosted the local economy. Most of them coming back to start their own businesses are equipped with the necessary information technology as well as many other urban skills, such as business consciousness, which enable them to re-feed the local farmers in the fields of culture, information, and technology to a certain extent [75]. They played an important catalytic role in the formation of Taobao villages.

A very typical case is Dr. Hu and his wife, Mrs. Meng. As the first group of young people in Daiji Town to return to their hometown to start businesses, they won the title of "Shandong e-commerce entrepreneurial leader." Their enterprise has grown into a well-known local leading enterprise in Hanfu, which integrates the industrial ecosystem of the design department, standardized production workshop, finished product display, original new product release, traditional e-commerce, and new media live e-commerce, driving the employment of more than 200 local villagers (EL1).

"In October 2014, when my husband and I went back to our hometown to visit relatives, we saw many people in the village engaged in e-commerce. Soon after, we decided to set up our own store on Taobao. In 2017, the "Hanfu craze" began to rise, which We thought was a good momentum of cultural consumption. In the second year, we quickly turned to the "battlefield" of Hanfu". From imitation to independent design, we set up several Hanfu stores on major e-commerce platforms and even cross-border e-commerce platforms (EL2)".

"I majored in fine arts as an undergraduate, so I feel very happy to engage in clothing design since I can not only make use of the professional knowledge I had learned, but also develop my own interest at the same time. And the other knowledge learned in the university is also a helper for me to do this, equipped with which I can learn things faster than other villagers, such as the layout design and the operation and maintenance of platform (EL2)".

Young entrepreneurs coming back to their hometowns call on more young people to start businesses in rural areas through their neighborhoods, peer groups, and other interpersonal networks. The reverse migration of the population drives the inflow of capital and information technology, which is a phenomenon of social revival promoted by the development of e-commerce. We saw this word in a vignette from the local government to young migrant workers:

"Your hometown has embracing you with ushering in young entrepreneurship boom. Running around outside is not good as running stores on Taobao, which is a sunrise industry with a prosperous future. Let's keep up with the pace of time, sit at home, tap the keyboard and play the movement of youth entrepreneurship (V3)".

There is a clear dual epistemology between urban and rural areas, in which the rural corresponds to a natural way of life while the urban relates to a realized center [76]. Digitalization has intensified the inextricable connection between urban and rural areas, which has enabled the modern lifestyle of the city to extend into the countryside. With the change in working patterns resulting from the transition from the traditional production mode, the villagers' concept of time also took a new turn. As office workers who have fixed working hours, most consumers buy goods during off-work hours or at night. In this circumstance, villagers who used to "work at sunrise and rest at sunset" are forced to change their timetable. In order to provide real-time online consulting services, they often have to work until midnight, making the distinction between family, leisure, and work blurred. As described by (V5) and observed by us, the landscape of urban life quietly emerges in Daiji Town:

"Around 4:00 p.m. is the time for the daily centralized delivery of online merchants and at that time, the crossroads in the center of the town are always jammed with cars, express trucks and fabric transport vehicles, which is exactly like the same "evening peak" in the big city (V5)".

"Restaurants springs up like bamboo shoots in the town. On summer nights there will be 20 or 30 barbecue stalls and midnight snack stalls, making the streets still brightly lit at 11 or 12:00 at night. It is hard to imagine that before the development of e-commerce, there were only 2 or 3 restaurants in Daiji Town and the streets were almost empty after 8:00 p.m. (GO4)".

4.3. Collaborative Governance

After being embedded in rural society, digitalization brought about the differentiation of internal interests in rural areas through far-reaching and extensive commercial processes, promoting the deconstruction and reconstruction of rural governance structures. In the context of the unique institutional conditions in rural China, special attention is paid to the interests and power relations of village organizations, villagers, and floating populations (including floating elites), as well as local governments in rural production and reproduction spaces [77]. Rural stakeholders build trust through dialogues, further, form a consensus commitment to the direction of rural development, and organize joint actions on the basis of existing institutional arrangements, village resources, and knowledge [78].

In the case of Daiji Town, the market increment dominated by performance costumes was still in its infancy. Being not sensitive to this kind of local change, all kinds of stakeholders did not have a definite risk perception or accurate market expectations. However, the extremely low entry threshold enabled the participants in the rural e-commerce industry to enter continuously and tentatively. And when e-commerce became the main source of income for rural households in Daiji Town, which coincided with the goal that the rural collective and government had been committed to promoting farmers to get rid of poverty and become rich, the governments began to realize the opportunities and benefits of rural e-commerce development. Therefore, building Taobao villages has been supported and promoted by the local governments as a way of rural revitalization. Eventually, the local governments, along with the rural collectives and industry associations, will play active roles in the decision-making process of rural governance, guided by the consensus to promote the development of e-commerce.

The local governments have the power to allocate resources and the organizational ability to mobilize urban and rural development, and they are the main institutional designers and key providers of starting conditions for rural governance. The government of Daiji Town has set up a Taobao Development Office, which is especially responsible for affairs related to the development of rural e-commerce (GO2). And they have made "Preferential Policies for Encouraging the Development of E-Commerce," which mainly include simplifying administrative procedures for registering e-commerce companies, appropriately reducing or exempting taxes, and providing financial incentives for e-commerce enterprises. They have also given large-scale financial investment to the infrastructure construction of Daiji Town, making its level higher than that of ordinary rural areas. Specifically, to meet the demand of tens of thousands of online operators for instantaneous electric power and network speed during the peak season, the governments have carried out power grid upgrades and fiber-to-the-home projects, which have improved the carrying capacity of rural information infrastructure. Moreover, to meet the logistics needs, they have also carried out a comprehensive transformation of the main roads in this town, including road resurfacing and widening (GO3).

"At first, we were very afraid of power failure and network outage. In this case, customers couldn't find and contact us when they need to. The conditions in our town were poor and sometimes the electric power would be cut off when it rained heavily. The network was also unstable, so our home not only had access to the Unicom Network, but also the Mobile Network and Telecom Network in case of the sudden network outage(V3)".

The governance affairs of rural collectives have changed from the traditional organization of irrigation, self-defense, dispute mediation, mutual assistance, entertainment, and clan activities to accelerate the development of e-commerce. The village collectives have assumed the dual identities of "decision-makers of the family" and "agents". Under the background of collective ownership of rural land in China, the collectively operated construction land is owned by rural collectives and can be used for rural production and business activities. In our case, the collective operational construction land in Sunzhuang Village, Dinglou Village, and other villages was jointly developed, providing sufficient space for the construction of public service agencies for e-commerce and a number of express logistics companies. In addition, the village collectives cooperated with the deployment of the government as organizers to carry out specific tasks to promote rural development, including organizing villagers to join in the construction of infrastructure such as roads and communications and conducting e-commerce skills training for villagers (VC1).

Driven by economic interests, the villagers' subjectivity awakened rapidly, and they began to actively participate in village affairs instead of being indifferent as in the past. Through their participation in voting, villagers elected the elites of e-commerce to take important positions in rural organizations and enter the core of rural governance as representatives of their own interests.

"In the transition election of the Village Party Branch Committee and Villagers' Committee of Dinglou Village in Daiji Town in October 2014, the villagers actively participated in the democratic vote. Finally, Ren Qingsheng, who was believed to do well in e-commerce and have the ability to lead the villagers to become rich through e-commerce, was elected as the village party secretary. And before engaged in e-commerce, he was just an electrician in the village and worked outside for a long time (GO3)".

With the surge of e-commerce merchants, bad competitive behaviors among enterprises began to appear in the market, such as underselling and placing orders maliciously to give moderate or bad reviews. The market began to exhibit bad competitive behavior among enterprises, such as selling at low prices, maliciously issuing orders to medium and poor ratings, and so on. In addition, due to poor tax supervision of rural e-commerce, many merchants are evading taxes. More professional organizations or groups were urgently needed to join the governance network. In this circumstance, an e-commerce industry association was established in Daiji Town in 2016, with a local e-commerce elite as its president and the performance costume processing enterprises and merchants on Taobao in this town as its members voluntarily. The e-commerce association strengthened coordination, cooperation, and exchange among e-commerce enterprises, shared industry market information, strengthened industry self-discipline, and enhanced the ability of enterprises to resist risks. Despite the many merits that e-commerce development brings, one caveat should be warranted: tax evasion is the dark side of e-commerce, and our case in Daiji Town could never eschew this challenge. Thus, the association can do little about fixing the tax evasion, nor can the local government. The e-commerce association standardized the market behavior of enterprises. Firstly, we will ensure the quality of products in Daiji Town by organizing technical seminars for different enterprises. Secondly, we will conduct propaganda and education for the e-commerce merchants and maintain market order through the communication of our association in order to avoid mutual price depression and ensure the benign development of the market (GO3).

4.4. Land-Use and Spatial Changes

The flows of capital, personnel, goods, and information affect the function and form of rural landscapes [79]. With the blooming development of rural tourism, rural e-commerce, and other industries, many villages are becoming multi-functional and post-industrial [80]. The rural space is constantly reconstructed in the process of creative destruction, which is characterized by functional mixing and land cover change. The case of Daiji Town shows that urban spatial organization patterns and land use layouts appear in rural areas and form rural urbanization landscapes because the rural non-agricultural production space has an increasingly important position and the rural low-density land use landscape also changed accordingly.

Under the current rural land regime in China, the land directly related to villagers includes homesteads and contracted land [81]. The land use nature of the contracted land is generally cultivated land, whose main function is crop planting, and it cannot be changed at will as it is protected by the basic farmland system. And the use and transformation of homesteads are more flexible; they can not only be used for self-built housing but also be applied for rental with the consent of the villagers. This provides clues for understanding the evolution logic of non-agricultural production space in Daiji Town, from family workshops and factory sheds along the street to modern industrial parks.

According to our interviews with villagers, in the initial stage of the development of e-commerce in Daiji Town, the family workshop was the general production mode for households in this town. Villagers used their own houses or built simple factory sheds in the homestead courtyard as the main production space for clothing production and sales (Figure 5). The family had formed an intergenerational division of labor production pattern composed of youth, middle age, and old age, among which the young couples were mainly responsible for clothing design, manufacture, and online shop operation, while middle-aged parents participated in the relatively simple links in the process of clothing manufacture and clothing packaging. As all the production links were completed in the villagers' own houses, a kind of production space roughly mixed with residential function was formed and scattered in the countryside, leaving great public security risks, especially fire.

(**a**) (**b**) (**c**)

Figure 5. The family workshop that mixed residential and productive functions (Source: photos taken during the fieldwork). (**a**) Villager houses, (**b**) Production space, (**c**) computer runs online shops.

In order to adapt to new forms of production and consumption, complex spatial transformations often take place in rural areas. The Dinglou and Sunzhuang Villages, located in the south of Daiji Town, serve as an excellent epitome of the spatial transformation of a traditional village into a Taobao village (Figure 6). The villagers and rural collectives have explored a spatial development model of "joint construction," in which 22 factory sheds have been built spontaneously on both sides of the main roads in the village, known as "Taobao Street". The funds and land for the street construction came from villagers, and the organization of the construction and allocation of the right to use production space were in the charge of rural collectives. The part along the street of these factory sheds is shops for goods display and offices for e-commerce customer service, and the rear is space for production and storage. A large number of e-commerce practitioners are gathered on both sides of the "Taobao Street", i.e., the Wansang Road, making various productive factors such as information, capital, technology, and so on interact rapidly here.

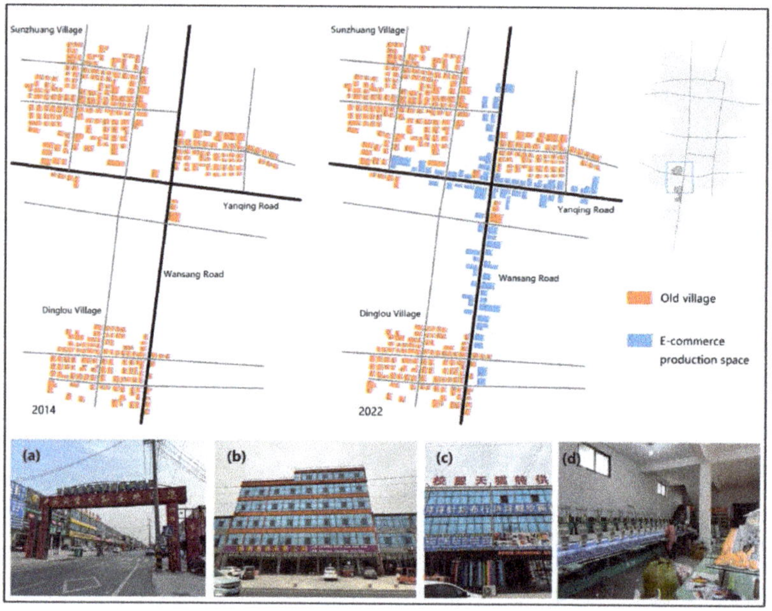

Figure 6. Spatial transformation of Dinglou and Sunzhuang village. (**a**) Wansang Road: known as the "Taobao Street"; (**b**,**c**): shops along the main street; (**d**): a small garment processing workshop for online shops (Source: author's mapping and photography).

However, the Taobao clusters built by villagers and rural collectives are proven to be informal. The property of informality is the result of rational economic strategy and "a spontaneous and creative response to the inability of the state to meet the basic needs of the poor" [82]. The land for the factory sheds comes from the contracted land of villagers, so the original agricultural production space is replaced by non-agricultural production space, which seems illegal in the context of China's national scheme to strictly protect the farmland. In addition, informality has also led to a series of safety and environmental issues in rural areas, challenging the government's governance capability.

With the continuous expansion of the e-commerce industry in Daiji town, the informal spatial expansion mode is proven to be unsustainable due to limited yearly land quotas and more strictly up-down environmental monitoring by China's central government. In 2017, the government allocated more than 300 mu of state-owned construction land for the construction of the Taobao industrial park in Daiji Town (Figure 7). According to our fieldwork, the industrial park was constructed in two phases, including the rural e-commerce integrated service building, 48 standardized factories of 1000 square meters, 220 shops, 18 logistics centers, and staff quarters and canteens. The supporting facilities and producer services of the industrial park were significantly attractive to new entrepreneurs and entrepreneurs who pursue high quality. It can be found that e-commerce has affected the spatial layout of Daiji town to a certain extent. With a more standardized and modern new layout, it can adapt to the new forms of production and consumption better.

(**a**) (**b**) (**c**) (**d**)

Figure 7. Taobao industrial park in Daiji Town (Source: photos taken during the fieldwork). (**a**) Industrial park entrance, (**b**) Standardized plant, (**c**) Express center, (**d**) modern supermarket.

5. Conclusions and Final Remarks

The era of digitization has blurred the boundary between urban and rural areas, reconfigured multiple elements and flows among them, and debased cities as the hegemonic centers of capital, talents, wealth, and information. It has enabled trans-border interactions between rural and urban areas and accelerated the urbanization process through intermingling four parts: the promotion of rural non-agricultural industries, the urbanization of farmers' identities and rural life, the innovation of rural governance structures, and spatializing rural areas. Thus, the highlight of this paper is to preemptively deliver an analytical framework that embodies the above four aspects of this new geo-economic phenomenon. Taking Daiji Town in China as an example, it analyzes the details and evidence of digitalization-enabled rural in-situ urbanization. The paper therefore extends the growing body of literature on rural in situ urbanization by synthesizing all interlinked elements instead of simplistic and homogenous accounts focused solely on rural industrialization. Through this integrative research, the main conclusive remarks and the main conclusions are as follows:

Firstly, digitalization connects the rural areas to the production and consumption networks in wider areas, promoting the transformation and prosperity of the rural economy. Wang et al. argue that the impact of e-commerce platforms such as Taobao on rural economic development varies greatly with the relative location of villages to major urban centers [60]. Our case study shows that digitalization can directly connect the scattered rural areas of China to modern society through the internet and logistics network, thus relieving them of the status of being isolated from cities caused by the limitations of local industry, transportation, or other factors and attached to the global urban pipeline network. Digital e-commerce has opened the network market, enabling the original market to extend from rural areas to the outside world and establishing a new network form of production and consumption. The internet platform realizes the rapid integration of online and offline industries, which contributes to the formation of large-scale non-agricultural industrial clusters in rural areas. And the large-scale development of industries further leads to the improvement of relevant service systems, which precisely lay the industrial foundation for urbanization. However, most of the "Taobao villages" developed from the bottom up are mainly engaged in traditional processing industries, with family workshops constituting the main production and operation units. This small-scale operation mode, which is mainly focused on low-end product processing, is not only spatially inefficient but also obtains limited value in the whole value chain.

Secondly, it should be highly noted that the reverse migration of young generations to rural areas becomes the mainstay dynamic of urbanization in the vast rural areas. The new employment opportunities created by e-commerce have greatly enhanced the attractiveness of rural areas to young people. As stated in our research, the reverse immigrants include not only the migrant workers who moved to big cities from the region earlier but also university graduates and even doctors. These well-educated immigrant entrepreneurs engage their knowledge and skills in the development of rural economies and actively participate in the governance of rural affairs. This could be conceptualized as a new mode of urbanization in China, which should be further advanced considering the population decline and aging, as well as the outmigration of educated people to rural areas.

Thirdly, digitization re-spatializes the vast rural areas. The rural space in various forms anchors digitalized e-commerce, whereby the development of e-commerce has reshaped the rural landscape at the same time. Thus, the land of farmers' own homesteads, as low-cost production space, is of significant assistance for the primitive accumulation of capital for e-commerce practitioners. When the spatial patterns of traditional rural areas are difficult to meet the needs of newly emerging e-commerce industrialization, the modernization and comprehensive expansion of rural space begin to lead to a significant process of spatial urbanization. By probing much deeper into the complex and yet intertwined spatiality of digitalized e-commerce, this emerging body of geographical research will gain new momentum in the context of worldwide disruptions and geopolitical rivalries.

Finally, the digitalization broaches new tracks to wipe off the urban-rural binary in the Global South, far from being epitomized by China's specificity. Nor is it to simplify the urbanization process by turning rural areas into cities through digitalized industrial growth; the goal of this paper is not to elevate digitization as a "one-size-fits-all" panacea to address rural plights. Instead, we hope to enrich and nuance rural studies beyond the extant scholarship to unpack the modernization of rural society by disrupting the urban-rural binary through the lens of digitalized urbanization. The conceptual schema developed in this paper is far from exhaustive; it is urgently needed that cross-fertilizing comparative case research among over 2000 Taobao towns in China be advanced. Further studies with variegated methodological reflexes and cross-disciplinary introspection are needed to enrich the spatiality and temporality of rural digitalized urbanization. As our case study shows, the relationship between urban and rural areas has been reconstructed due to the changing patterns of production and consumption through digitalized e-commerce. When the demand for employment and public services can be met locally in rural areas, it might be the only choice for rural outmigrants to migrate into metropolitan areas or cities. It merits being noted that the self-induced rural construction movement of Chinese farmers against the background of digitalization is a bottom-up urbanization process. However, this bottom-up urbanization has also faced some challenges. Due to the insufficient space capacity of rural areas for this possibility, the phenomenon of empty rural housing waste and the shortage of e-commerce production space coexist in Taobao villages, which need effective governance responses from the innovation of rural land use systems and the compilation and implementation of local rural planning.

Author Contributions: Conceptualization, L.L. and T.S.; Methodology, L.L.; Software, L.L.; Writing—original draft, L.L.; Writing—review & editing, T.S. All authors have read and agreed to the published version of the manuscript.

Funding: This study was funded by National Social Science Foundation of China [No.19CJY015] and National Natural Science Foundation of China [No.42171180].

Data Availability Statement: The data are sourced from fieldwork by authors.

Conflicts of Interest: The authors declare no conflict of interest.

References

1. Roberts, E.; Anderson, B.A.; Skerratt, S.; Farrington, F. A review of the rural-digital policy agenda from a community resilience perspective. *J. Rural Stud.* **2017**, *54*, 372–385. [CrossRef]
2. Peng, C.; Ma, B.; Zhang, C. Poverty alleviation through e-commerce: Village involvement and demonstration policies in rural China. *J. Integr. Agric.* **2021**, *20*, 998–1011. [CrossRef]
3. Karine, H. E-commerce development in rural and remote areas of BRICS Countries. *J. Integr. Agric.* **2021**, *20*, 979–997.

4. Zhou, J.; Yu, L.; Choguill, C.L. Co-evolution of technology and rural society: The blossoming of taobao villages in the information era, China. *J. Rural Stud.* **2021**, *83*, 81–87. [CrossRef]
5. Wang, Y. Originally, You Are Such a Cao County. *Science and Technology Daily*, 19 May 2021. Available online: https://m.thepaper.cn/baijiahao_12749179 (accessed on 2 March 2023). (In Chinese)
6. Administration and Management Institute; MARA; Ali Research. *Study on Agricultural Products E-Commerce out of Villages and into Cities: Taking Ali Platform as an Example*; Administration and Management Institute: Wotton-Under-Edge, UK, 2021. (In Chinese)
7. Ali Research, 2018. China Taobao Village Development Report (2014–2018). Available online: https://www.sohu.com/a/283404163_775264 (accessed on 6 January 2023). (In Chinese)
8. Ali Research, 2019. China Taobao Village Development Report (2009–2019). Available online: https://www.sohu.com/a/339624966_407401 (accessed on 4 January 2023). (In Chinese)
9. Zuo, C. The List of Taobao Villages in 2021 Has been Released and the Number of Taobao Villages in China Has Exceeded 7000. Ali Research, 12 October 2021. Available online: http://www.aliresearch.com/ch/information/informationdetails?articleCode=256317657652006912&type=%E6%96%B0%E9%97%BB (accessed on 13 March 2023). (In Chinese)
10. Zhang, Y.; Long, H.; Ma, L.; Tu, S.; Li, Y.; Ge, D. Analysis of rural economic restructuring driven by e-commerce based on the space of flows: The case of Xiaying village in central China. *J. Rural Stud.* **2022**, *93*, 196–209. [CrossRef]
11. Qiu, Z. How e-commerce Reshapes the Relationship among Actors in Rural Society: A New Political, Economic and Social Triangle Perspective. *J. Chin. Acad. Gov.* **2018**, *1*, 47–54+149. (In Chinese)
12. Lin, G.; Xie, X.; Lv, Z. Taobao practices, everyday life and emerging hybrid rurality in contemporary China. *J. Rural Stud.* **2016**, *47*, 514–523. [CrossRef]
13. Tang, W.; Zhu, J. Informality and Rural Industry:Rethinking the Impacts of E-Commerce on Rural Development in China. *J. Rural. Stud.* **2020**, *75*, 20–29. [CrossRef]
14. Liu, M.; Zhang, Q.; Gao, S.; Huang, J. The spatial ggregation of rural e-commerce in China: An empirical investigation into Taobao Villages. *J. Rural Stud.* **2020**, *80*, 403–417. [CrossRef]
15. Ma, H.; Li, Q.; Liu, J.; Guo, X.; Liu, M.; Wang, F. Spatial Pattern Characteristics and Influencing Factors of Taobao Town in China. *Econ. Geogr.* **2017**, *37*, 118–124. (In Chinese)
16. Chen, H.; Zhang, J. Taobao Villages: Rural development transformation driven by the space of flow. *City Plan. Rev.* **2018**, *42*, 97–105. (In Chinese)
17. Parra-López, C.; Reina-Usuga, L.; Carmona-Torres, C.; Sayadi, S.; Klerkx, L. Digital transformation of the agrifood system: Quantifying the conditioning factors to inform policy planning in the olive sector. *Land Use Policy* **2021**, *108*, 105537. [CrossRef]
18. Zhang, G. How does Internet Mechanism activate rural Economy-Based on the Study of Caoxian County in Shandong Province and Suining City in Jiangsu Province. *Dev. Small CitiesTowns* **2016**, *6*, 16–17. (In Chinese)
19. Zhan, J.; Lu, C. Study on the Influence of Digital Economy on Rural E-commerce in China. *World Surv. Res.* **2022**, *9*, 3–11. (In Chinese)
20. Cohen, B. Urbanization in developing countries: Current trends, future projections, and key challenges for sustainability. *Technol. Soc.* **2006**, *28*, 63–80. [CrossRef]
21. Shatkin, G. Global cities of the South: Emerging perspectives on growth and inequality. *Cities* **2007**, *24*, 1–15. [CrossRef]
22. Brenner, N.; Schmid, C. Towards a new epistemology of the urban? *City Anal. Urban Trends Cult. Theory Policy Action* **2015**, *19*, 151–182. [CrossRef]
23. Jain, M.; Korzhenevych, A. The concept of planetary urbanization applied to India's rural to urban transformation. *Habitat Int.* **2022**, *129*, 102671. [CrossRef]
24. Gambe, T.R.; Turok, I.; Visagie, J. The trajectories of urbanisation in Southern Africa: A comparative analysis. *Habitat Int.* **2023**, *132*, 102747. [CrossRef]
25. Bhagat, R.B.; Mohanty, S. Emerging pattern of urbanization and the contribution of migration in urban growth in India. *Asian Popul. Stud.* **2009**, *5*, 5–20. [CrossRef]
26. Zhang, K.H.; Song, S. Rural-urban migration and urbanization in China: Evidence from time-series and cross-section analyses. *China Econ. Rev.* **2003**, *14*, 386–400. [CrossRef]
27. Zhou, T.; Jiang, G.; Zhang, R.; Zheng, Q.; Ma, W.; Zhao, Q.; Li, Y. Addressing the rural in situ urbanization (RISU) in the Beijing-Tianjin-Hebei region: Spatio-temporal pattern and driving mechanism. *Cities* **2018**, *75*, 59–71. [CrossRef]
28. Lucas, R.E.B. *Internal Migration and Urbanization: Recent Contributions and New Evidence*; Institute for Economic Development, Boston University: Boston, MA, USA, 1998.
29. Xing, C.; Zhang, J. The preference for larger cities in China: Evidence from rural-urban migrants. *China Econ. Rev.* **2017**, *43*, 72–90. [CrossRef]
30. Siciliano, G. Urbanization strategies, rural development and land use changes in China: A multiple-level integrated assessment. *Land Use Policy* **2012**, *29*, 165–178. [CrossRef]
31. Long, H.; Zou, J.; Liu, Y. Differentiation of rural development driven by industrialization and urbanization in eastern coastal China. *Habitat Int.* **2009**, *33*, 454–462. [CrossRef]
32. Liu, Y.; Liu, Y.; Chen, Y.; Long, H. The process and driving forces of rural hollowing in China under rapid urbanization. *J. Geogr. Sci.* **2010**, *20*, 876–888. [CrossRef]
33. Mughal, M.A.Z. Rural urbanization, land, and agriculture in Pakistan. *Asian Geogr.* **2019**, *36*, 81–91. [CrossRef]

34. Ma, L.J.C.; Fan, M. Urbanisation from Below: The Growth of Towns in Jiangsu, China. *Urban Stud.* **1994**, *31*, 1625–1645. [CrossRef]
35. Zhu, Y. In Situ urbanization in rural China: Case studies from Fujian province. *Dev. Chang.* **2000**, *31*, 413–434. [CrossRef]
36. Zhu, Y. *Changing Urbanization Processes and In Situ Rural-Urban Transformation: Reflections on China's Settlement Definitions. New Forms of Urbanization—Beyond the Urban-Rural Dichotomy Edited by T. Champion and G. Hugo*; Ashgate Publishing: Aldershot, UK, 2004.
37. Shen, X.; Ma, L.J.C. Privatization of rural industry and de facto urbanization from below in southern Jiangsu, China. *Geoforum* **2005**, *36*, 761–777. [CrossRef]
38. Zhu, Y.; Wang, W.W.; Lin, L.; Shen, J.; Ren, Q. Return migration and in situ urbanization of migrant sending areas: Insights from a survey of seven provinces in China. *Cities* **2021**, *115*, 103242. [CrossRef]
39. Li, L.; Li, X.; Hai, B.; Wang, X.; Xu, J. Evolution of rural settlement in an inland nonmetropolitan region of China at a time of rapid urbanisation: The case of Gongyi. *J. Rural Stud.* **2020**, *79*, 45–56. [CrossRef]
40. Irwin, E.G.; Isserman, A.M.; Kilkenny, M.; Partridge, M.D. A century of research on rural development and regional issues. *Am. J. Agric. Econ.* **2010**, *92*, 522–553. [CrossRef]
41. Liu, S.; Wang, Y. From Native Rural China to Urban-rural China: The Rural Transition Perspective of China Transformation. *Manag. World* **2018**, *34*, 128–146+232. (In Chinese)
42. Long, H.; Liu, Y.; Li, X.; Chen, Y. Building new countryside in China: A geographical perspective. *Land Use Policy* **2010**, *27*, 457–470. [CrossRef]
43. Long, H.; Liu, Y. Rural restructuring in China. *J. Rural Stud.* **2016**, *47*, 387–391. [CrossRef]
44. Salemink, K.; Strijker, D.; Bosworth, G. Rural development in the digital age: A systematic literature review on unequal ICT availability, adoption, and use in rural areas. *J. Rural Stud.* **2017**, *54*, 360–371. [CrossRef]
45. Hoq, K.M.G. Role of information for rural development in Bangladesh: A sector-wise review. *Inf. Dev.* **2012**, *28*, 13–21. [CrossRef]
46. Murdoch, J. Networks—A new paradigm of rural development? *J. Rural Stud.* **2000**, *16*, 407–419. [CrossRef]
47. Luo, Z.; He, H. New Urbanization Form Below in China: Rural Urbanization Driven by E-commerce. *City Plan. Rev.* **2017**, *41*, 31–40. (In Chinese)
48. Lou, J.; Hu, D. Taobao Village, real-time urbanization and new urbanization practices. *Acad. Res* **2018**, *5*, 58–62. (In Chinese)
49. Lin, Y. E-urbanism: E-commerce, migration, and the transformation of Taobao villages in urban China. *Cities* **2019**, *91*, 202–212. [CrossRef]
50. Tang, Y.; Chen, M. The Impact of Agricultural Digitization on the High-Quality Development of Agriculture: An Empirical Test Based on Provincial Panel Data. *Land* **2022**, *11*, 2152. [CrossRef]
51. Xu, N.; Zhao, D.; Zhang, W.; Liu, M.; Zhang, H. Does Digital Transformation Promote Agricultural Carbon Productivity in China? *Land* **2022**, *11*, 1966. [CrossRef]
52. Zhao, S.X.B.; Wong, K.K.K. The sustainability dilemma of China's township and village enterprises: An analysis from spatial and functional perspectives. *J. Rural Stud.* **2002**, *18*, 257–273. [CrossRef]
53. Han, W.; Wei, Y.; Cai, J.; Yu, Y.; Chen, F. Rural nonfarm sector and rural residents' income research in China. An empirical study on the township and village enterprises after ownership reform (2000–2013). *J. Rural Stud.* **2021**, *82*, 161–175. [CrossRef]
54. Thomä, J. An urban-rural divide (or not?): Small firm location and the use of digital technologies. *J. Rural Stud.* **2023**, *97*, 214–223. [CrossRef]
55. Martina, A. One size does not fit all: The plurality of knowledge sources for transition to sustainable farming. *J. Rural Stud.* **2023**, *97*, 243–254.
56. Rose, D.C.; Sutherland, W.J.; Barnes, A.P.; Borthwick, F.; Ffoulkes, C.; Hall, C.; Moorby, J.M.; Nicholas-Davies, P.; Twining, S.; Dicks, L.V. Integrated farm management for sustainable agriculture: Lessons for knowledge exchange and policy. *Land Use Policy* **2019**, *81*, 834–842. [CrossRef]
57. Hoogendoorn, G.; Visser, G. The role of second homes in local economic development in five small South African towns. *Dev. South. Afr.* **2010**, *27*, 547–562. [CrossRef]
58. Wu, M.; Gallent, N. Second homes, amenity-led change and consumption-driven rural restructuring: The case of Xingfu village, China. *J. Rural Stud.* **2021**, *82*, 391–403. [CrossRef]
59. Woods, M. Engaging the global countryside: Globalization, hybridity and the reconstitution of rural place. *Prog. Hum. Geogr.* **2007**, *31*, 485–507. [CrossRef]
60. Wang, C.; Miao, J.T.; Phelps, N.A.; Zhang, J. E-commerce and the transformation of the rural: The Taobao village phenomenon in Zhejiang Province, China. *J. Rural Stud.* **2021**, *81*, 159–169. [CrossRef]
61. Rosen, J.; León, L.F.A. The Digital Growth Machine: Urban Change and the Ideology of Technology. *Ann. Am. Assoc. Geogr.* **2022**, *112*, 2248–2265. [CrossRef]
62. Sousa, M.J. Digital Technologies and Public Policies Applied to Green Cities. *Land* **2022**, *11*, 2097. [CrossRef]
63. European Commission. Smart Cities—Smart Living. 2019. Available online: https://ec.europa.eu/digital-single-market/en/smart-cities (accessed on 11 March 2023).
64. Navío-Marco, J.; Rodrigo-Moya, B.; Gerli, P. The rising importance of the "Smart territory" concept: Definition and implications. *Land Use Policy* **2020**, *99*, 105003. [CrossRef]
65. Liang, L.; Li, Y. How does government support promote digital economy development in China? The mediating role of regional innovation ecosystem resilience. *Technol. Forecast. Soc. Chang.* **2023**, *188*, 122328. [CrossRef]

66. Jamil, S. From digital divide to digital inclusion: Challenges for wide-ranging digitalization in Pakistan. *Telecommun. Policy* **2021**, *45*, 102206. [CrossRef]
67. OECD. *Understanding the Digital Divide*; OECD: Paris, France, 2001.
68. D'Amico, G.; Arbolino, R.; Shi, L.; Yigitcanlar, T.; Ioppolo, G. Digitalisation driven urban metabolism circularity: A review and analysis of circular city initiatives. *Land Use Policy* **2022**, *112*, 105819. [CrossRef]
69. Tian, Q.; Guo, L.; Zheng, L. Urbanization and rural livelihoods: A case study from Jiangxi Province, China. *J. Rural Stud.* **2016**, *47*, 577–587. [CrossRef]
70. Chen, X.; Zhang, X. Spatial Layout and Industrial Development of E-commerce Characteristic Towns- A Case Study of Taobao-towns. *Forum Sci. Technol. China* **2018**, *6*, 91–97+120. (In Chinese)
71. Fei, X. *From the Soil: The Foundations of Chinese Society*; Observatory Press: Shanghai, China, 1948. (In Chinese)
72. Liu, Y. Research on the urban-rural integration and rural revitalization in the new era in China. *Acta Geogr. Sin* **2018**, *73*, 637–650. (In Chinese)
73. Jiang, Q.; Li, Y.; Si, H. Digital Economy Development and the Urban–Rural Income Gap: Intensifying or Reducing. *Land* **2022**, *11*, 1980. [CrossRef]
74. Qi, J.; Zheng, X.; Guo, H. The formation of Taobao villages in China. *China Econ. Rev.* **2019**, *53*, 106–127. [CrossRef]
75. Chen, R. Self-consciousness and Rebuilding of Rural Areas: Social Analysis of the Phenomenon of Taobao Village. *J. Huazhong Agric. Univ. Soc. Sci. Ed.* **2016**, *3*, 74–81. (In Chinese)
76. Williams, R. *The Country and the City*; Oxford University Press: New York, NY, USA, 1975.
77. Lai, Y.; Chan, E.H.W.; Choy, L. Village-led land development under state-led institutional arrangements in urbanizing China: The case of Shenzhen. *Urban Stud.* **2017**, *54*, 1736–1759. [CrossRef]
78. Zhang, Q.; Ye, C.; Duan, J. Multi-dimensional superposition: Rural collaborative governance in Liushe Village, Suzhou City. *J. Rural Stud.* **2022**, *96*, 141–153. [CrossRef]
79. Kizos, T.; Primdahl, J.; Kristensen, L.S.; Busck, A.G. Introduction: Landscape Change and Rural Development. *Landsc. Res.* **2010**, *35*, 571–576. [CrossRef]
80. Liu, B.; Fang, Y. Spatial Evolution and Optimization of "Taobao" Village E-commerce Production Space: A Case Study of Dinglou Village in Caoxian County of Shandong Province. *Areal Res. Dev.* **2020**, *39*, 138–144. (In Chinese)
81. Li, Y.; Liu, Y.; Long, H.; Cui, W. Community-based rural residential land consolidation and allocation can help to revitalize hollowed villages in traditional agricultural areas of China: Evidence from Dancheng County, Henan Province. *Land Use Policy* **2014**, *39*, 188–198. [CrossRef]
82. Rakowski, C.A. *Contrapunto: The Informal Sector Debate in Latin America*; SUNY Press: Albany, NY, USA, 1994.

land

MDPI

Article

Uncovering the Triple Synergy of New-Type Urbanization, Greening and Digitalization in China

Qiangyi Li [1,2], Jiexiao Ge [1,*], Mingyu Huang [1], Xiaoyu Wu [1] and Houbao Fan [3,*]

[1] School of Economics and Management, Guangxi Normal University, Guilin 541006, China
[2] Key Laboratory of Digital Empowerment Economic Development, Guangxi Normal University, Education Department of Guangxi Zhuang Autonomous Region, Guilin 541006, China
[3] School of Applied Economics, Jiangxi University of Finance and Economics, Nanchang 330013, China
* Correspondence: gejiexiao@stu.gxnu.edu.cn (J.G.); 2202226587@stu.jxufe.edu.cn (H.F.)

Abstract: The in-depth discussion and analysis of the synergistic effect of new-type urbanization, greening and digitalization (NUGD) is important for the achievement of sustainable social, ecological and economic development. Therefore, in this study, an evaluation index system composed of these three subsystems was constructed for Chinese cities from 2011 to 2021. The comprehensive and collaborative development levels of each subsystem were measured by means including the entropy weight method and the coupling coordination model, respectively. Then, methods such as ESDA and the Dagum Gini coefficient were applied to investigate the spatiotemporal evolution and spatial differences in the triple synergy effect of the NUGD system in Chinese cities. Finally, the constraining factors of the triple synergy effect were revealed using the obstacle degree model. The findings demonstrated the following: (1) Overall, the NUGD subsystems and their comprehensive levels were increasing, with moderate overall development levels. (2) The synergistic development of the NUGD system exhibited an upward trend. Spatially, the synergistic development level showed distinct differentiation, being higher in the east and lower in the west. The multidimensional dynamic variation characteristics obtained through kernel density estimation revealed that the triple synergy level exhibits high stability. (3) The differences within the east and between the eastern and western areas were the largest, with the intensity of transvariation as the main source. (4) The five criterion layers, including social and spatial urbanization variables, were the key constraints that affected the triple synergy of the NUGD in Chinese cities, and the restrictive role of factors such as the proportion of urban construction land and the per capita postal business volume should not be ignored. This study provides a valuable reference and decision-making guidance to promote China's acceleration toward a new urbanization path supported by both digitalization and green transformation.

Keywords: triple synergy of the NUGD system; spatiotemporal evolution; dynamic progression; obstacle factors

1. Introduction

Currently, the most prominent features of global economic and social development are urbanization, greening and digitalization [1]. As early as the beginning of the 21st century, significant progress was achieved in the global urbanization process, with cities reaching high levels in terms of scale expansion, coverage and diversified development [2].Moreover, it is expected that, by 2070, global population will reach 9.7 billion, and two-thirds of the population will have flowed into urban areas [3]. It can be observed that, in a broad sense, urbanization is defined based on population changes. For example, in some Latin American countries, an area with a population of 2500 is considered urban; in other countries, the threshold is 5000. Meanwhile, in India, areas accommodating 100,000 residents are considered urban areas [4]. In fact, benefiting from the reform and opening up process, the urbanization process in China has shown unprecedented growth, with the urbanization rate increasing from 17.92% in 1978 to 64.72% in 2021 [5]. However, rapid urbanization has a profound impact on climatic conditions, vegetation, lake ecosystems and global land resources [6–8]; furthermore, existing studies have shown that urbanization causes extensive and permanent damage to river systems [9]. In addition, the rapid expansion of cities and population growth have further reshaped the urban landscape and reduced urban biodiversity [10]. To address this challenge, which may be a common issue faced by most countries, this study takes China as an example and introduces the concepts of green development and digital transformation, emphasizing the importance of greening and digital empowerment and highlighting the urgency of sustainable urban development. Furthermore, from a sustainable development perspective, urbanization development should shift toward a new type—that is, new-type urbanization. Chinese new-type urbanization places a greater emphasis on promoting the citizenship of agricultural migrant workers in an orderly manner, ensuring that they have access to basic public services in towns and highlighting its human-oriented and pro-urban–rural equality characteristics [11,12]. In addition, it emphasizes optimizing the urban layout and spatial structure and places an emphasis on enhancing the sustainable development capacity and quality of urban development. It comprehensively considers various factors, such as the population, economic growth, social welfare and the spatial structure, aiming to guide cities toward inclusive, safe, resilient and sustainable development [13–15].

Promoting the in-depth integration of human-centered new-type urbanization construction and greening is essential to realize the goal of environmental sustainability and harmonious coexistence between humans and nature [16,17]. With the frequent occurrence of global natural disasters, rising sea levels and ecosystem destruction becoming the main obstacles constraining sustainable development, green development has been increasingly adopted in many countries to address challenges such as resource scarcity and the adverse effects of climate change [18,19]. For example, the UK has announced the Low-Carbon Transformation Plan and the Renewable Energy Strategy; in 2019, the European Union issued the "European Green Deal", proposing to become the world's first carbon-neutral country by 2050, expressing its pursuit of green and low-carbon development [20]. Moreover, in 2020, the dual-carbon goals were proposed by the Chinese government, and subsequently, a series of green and low-carbon transformation work plans were implemented to strengthen its ecological environmental protection and achieve its sustainable development goals, clearly demonstrating China's responsibility as a major country [21,22]. Greening is the path to realizing green development, and it includes the transformation of multi-dimensional development models such as economic–social–ecological models and achieving green production and industries at the economic level, green life at the social level and green protection and governance at the ecological level [23]. Its purpose is to reduce the pressure of economic and social development on the ecological environment, achieve efficient outputs and advocate for a green and environmentally friendly way of life. By implementing effective environmental governance measures, it promotes society's transition toward sustainable development and environmental friendliness, fostering harmony between humans and nature [24]. Thus, it plays a profound role in improving the quality of urban development and building a beautiful China [25].

In the green transformation process, the role of digital empowerment has become increasingly prominent [26,27]. Following the 2015 Paris Agreement, countries worldwide have been working together to promote environmental protection and have reached a consensus on green transformation [28]. Notably, with the advent of the digital age, global economic, social and environmental governance have achieved significant results against the backdrop of digital transformation. For example, digital transformation provides many new employment opportunities, significantly changing the prospects for global economic development [29]. In addition, by transforming operation mode, productivity and production relationships, optimizing resource allocation and improving innovation efficiency, digitalization has played an increasing role in facilitating urban development at a high level and driving the transformation of the development mode to a green and low-carbon one [30,31]. Thus, in China, national policy documents and government work reports have repeatedly emphasized that digitalization is a significant engine for the improvement and growth of Chinese economic development. Through the dual role of technological change and policy support, it has been promoted as a key force for the long-term and sustainable development of the Chinese economy and society [32,33].

Importantly, the synergistic analysis in this study adopts a systems perspective, focusing on the dynamic changes in new-type urbanization, greening and digitalization while also revealing the complex interdependencies and interactions among the three major systems. New-type urbanization, greening and digitalization are important components of the complex social–ecological–economic system, and the promotion of the synergistic development of these three factors is a crucial step in addressing the common challenges faced by many countries around the world [34,35]. Unfortunately, there are currently few studies focusing on the dynamic interaction between these three systems. Therefore, referring to the relevant literature [36,37], this study places Chinese urban new-type urbanization, green development and digital transformation within the same research framework. Utilizing the coupling coordination model, ESDA and obstacle analysis, the following three objectives are determined: (1) to measure the levels of the three major subsystems and the NUGD system separately; (2) to describe the dynamic temporal and spatial distributions and regional differences in NUGD; and (3) to reveal the main obstacle factors affecting NUGD. In addition, the examination of this system combines different geographical, economic and social aspects. This study not only provides more targeted policy insights for sustainable development in China but also offers a template for sustainable development in other countries around the world that are facing similar challenges.

Therefore, in this study, 282 cities in China were adopted to explore the triple synergistic effect. The possible contributions herein may be as follows: (1) Regarding the research perspective, against the background of achieving long-term and sustainable socioeconomic development, this study focused on the inherent integration mechanisms and patterns of new-type urbanization, green transformation and digitalization. (2) Regarding the research content, the synergistic development of the NUGD system in Chinese cities was measured first, after which the spatial agglomeration and distribution characteristics were determined via the ESDA. Furthermore, regional differences and sources were analyzed. Finally, the factors affecting the improvement in the triple synergy level at the criterion and indicator levels were determined. (3) In terms of policy significance, the conclusions of this study enrich the related research on new-type urbanization, green transformation and digital development, providing a useful reference for the promotion of the steady progress of social sustainable development, ecological civilization construction and healthy economic growth in China and other countries.

2. Materials and Methods

2.1. Analysis of the Synergistic Development Mechanism of the NUGD System

Placing new-type urbanization, green transformation and digitalization within the same research framework and systematically analyzing the dynamic interactions among these three subsystems is of strategic importance in achieving new advancements in social civilization, improving the eco-quality and accelerating economic growth [38,39]. The three components of the NUGD system are highly interrelated and consistent, with its core being human-oriented, adhering to the unique path of digitalization and intelligence, promoting urban and rural development to a high level and steadily progressing toward the achievement of common prosperity [39,40]. Therefore, elucidating the internal mechanisms of the triple synergy of the NUGD system (Figure 1) is important in promoting the simultaneous enhancement of its social, ecological and economic benefits.

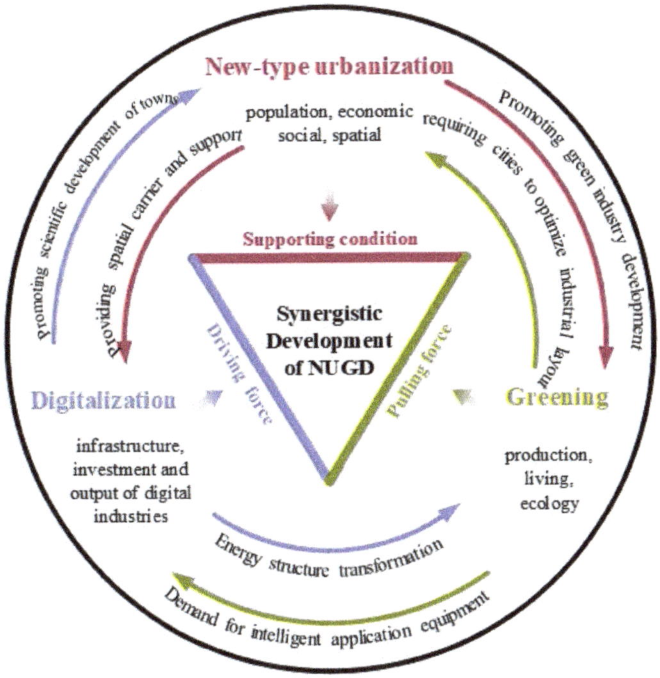

Figure 1. Synergistic development mechanism of the NUGD system.

Firstly, new-type urbanization serves as a supporting condition for the NUGD system. Chinese new-type urbanization is human-centered urbanization, comprehensively enhancing the quality of cities and optimizing the spatial layout [36]. Cities act as carriers and platforms for population aggregation, industrial transformation, information development and the provision of social services. Specifically, new-type urbanization provides a spatial carrier and application support for the digital economy. At the same time, new-type urbanization adheres to a development path characterized by intensity, intelligence, greenness and low carbon [41] and promotes environmentally friendly cities through the improvement of green infrastructure construction [42,43], the promotion of green industry development [40] and technological innovation [44].

Secondly, greening is the pulling force of the NUGD system. The Chinese government has stipulated that China should vigorously promote green development and foster a harmonious coexistence between humanity and nature. On one hand, although digitalization-assisted greening has great potential, the process of digital development itself is characterized by the notable issues of energy consumption and waste generation [45,46]. Greening imposes higher requirements and diverse demands on network transmission performance and intelligent application equipment. Thus, greening plays a leading role in driving digital transformation. On the other hand, as urbanization levels rise, negative issues such as environmental pollution and ecological damage are inevitable. Therefore, embedding the concept of green and low-carbon development into urban planning is particularly important. This is because green development can optimize the industrial layout and adjust the land use structure to continuously promote low-carbon development.

Thirdly, the development of digitalization is the driving force behind the NUGD system. Chinese national policy documents and government work reports emphasize the acceleration of digital development, seeking to exploit the full advantages of digital development, making it an important force for the continuous optimization and positive transformation of the Chinese economy and society [32,47]. On one hand, the digital economy is a new economic concept in urbanization, becoming a new engine for promoting high-quality development, improving resource allocation efficiency and transforming development methods [48]. On the other hand, the White Paper on the Collaborative Development of Digitalization and Greening (2022) (http://www.caict.ac.cn/english/research/whitepapers/202303/P020230316598975246514.pdf (accessed on 20 June 2024)) points out that digitalization-empowered greening and greening-assisted digitalization are mutually supportive, and they are crucial for the long-term development of the economy. The development of digitalization is the primary means of promoting a green transition [27]. Specifically, digitalization provides full-chain support for green development, promoting new models of low-carbon production and leading new trends in green living.

In summary, new-type urbanization provides supporting conditions for greening and digitalization, with greening pulling digitalization and urban development toward agglomeration, greenness and high efficiency. Digitalization empowers greening and new-type urbanization. The organic integration of these three concepts can accelerate the realization of a more prosperous, democratic, harmonious and beautiful China.

2.2. Construction of an Indicator System

The NUGD system is a comprehensive system that consists of the triple synergy of new-type urbanization, greening and digitalization. Based on existing research findings, an evaluation index system consisting of these three subsystems was established to measure their synergistic effect (Table 1). Differing from the traditional type, new-type urbanization is a human-oriented one, which is a multidimensional process involving the population, economy, society and space [12,49]. Among these aspects, population urbanization is its core feature, reflecting the scale and density of the population; economic urbanization promotes the agglomeration of production factors toward cities, with economic prosperity and income growth as its key characteristics; social urbanization aims to promote the equalization of basic public services, enhancing people's happiness and satisfaction; and spatial urbanization reflects the urban spatial layout, land development and utilization and their impacts on population movement and economic growth. Green development focuses on improving resource utilization efficiency and alleviating environmental pressure during the production process, enhancing residents' awareness of green environmental protection in their daily lives, improving waste disposal efficiency and promoting the systematic coordination of society, ecology and economy by simultaneously focusing on regional ecological construction and environmental protection [23,50]. In the dimension of digitalization, considering existing findings [51,52], we focused on analyzing the following three aspects: digital infrastructure, investment in digital industries and the output of digital industries. The Internet and mobile phones have provided digital resources and modern information networks for the development of digitalization, which are basic conditions. Investment in digital industries is the key to digital development, and in this study, it was measured in terms of digital practitioners, scientific and technological expenditure and digital talent resources. The output of digital industries is a core aspect of digital development, and the per capita telecommunications output and per capita postal output were regarded as relevant indicators.

Table 1. Indicators for synergistic development of NUGD.

Subsystem	Dimension	Index	Property
The Development of New-type Urbanization	Population urbanization B_1	Ration of persons engaged in secondary and tertiary industries X_1	+
		Registered unemployment rate X_2	−
		Population density X_3	+
		Urbanization rate X_4	+

Table 1. *Cont.*

Subsystem	Dimension	Index	Property
The Development of New-type Urbanization	Economic urbanization B_2	Per capita disposable income of urban residents X_5	+
		Ration of output value of secondary and tertiary industries X_6	+
		Proportion of local budgetary revenue X_7	+
	Social urbanization B_3	Number of doctors per 10^3 persons X_8	+
		Amount of hospitals beds per 10^4 persons X_9	+
		Per capita expenditure on education X_{10}	+
		Number of books per 100 people in libraries X_{11}	+
		the proportion of college teachers X_{12}	+
	Spatial urbanization B_4	Per capita road area X_{13}	+
		Ration of urban construction land X_{14}	+
		Ration of built-up area X_{15}	+
The Development of Greening	Green Production B_5	Centralized treatment rate of wastewater X_{16}	+
		Industrial wastewater discharge per unit of GDP X_{17}	−
		Industrial SO2 emissions per unit of GDP X_{18}	−
	Green Living B_6	Harmless treatment rate of domestic garbage X_{19}	+
		Number of public transport vehicles per 10^4 population X_{20}	+
		Ratio of residential land area to green space area X_{21}	−
		Per capita daily domestic water consumption X_{22}	−
	Green Ecology B_7	Proportion of green space area X_{23}	+
		Per capita park green area X_{24}	+
		Green coverage rate in built-up area X_{25}	+
The Development of Digitalization	Digital Infrastructure B_8	Internet penetration rate X_{26}	+
		Mobile phone penetration rate X_{27}	+
		Fiber optic cable density X_{28}	+
		Per capita GDP X_{29}	+
		Digital Financial Inclusion Index X_{30}	+
	Investment in digital industries B_9	Proportion of employed personnel in computer services and software industry X_{31}	+
		Proportion of science and technology expenditure X_{32}	+
		Number of students in higher education X_{33}	+
	Output of digital industries B_{10}	Telecommunications services per capita X_{34}	+
		Per capita total post X_{35}	+

2.3. Methods

2.3.1. Entropy Weight Method

Due to its ability to effectively overcome the drawbacks of subjective weighting methods, the entropy weight method is commonly used to determine weights based on information entropy. Therefore, this method was applied to measure the development levels of the three subsystems in this study. The detailed calculation process is shown below [2,16].

(1) Calculate the weight of indicator j:

$$W_j = \frac{d_j}{\sum_{j=1}^{n} d_j} = \frac{1 - e_j}{n - \sum_{j=1}^{n} e_j} \quad (1)$$

(2) Calculate the comprehensive index:

$$W_{ij} = \sum_{i=1}^{n} W_j \times y_{ij} \quad (2)$$

where y_{ij} is the standardized value of the indicator, d_j is the difference coefficient, W_j is the weight and W_{ij} is the comprehensive index.

2.3.2. Coupling Coordination Model

The extent of the interdependence and mutual restraint between multiple systems is reflected by the coupling level, while the degree of coordination indicates the beneficial coupling within the coupling relationships. The following process was applied to measure the NUGD system's synergistic development level [16,53]:

$$C = 3 \times \left[\frac{U_n U_g U_d}{(U_n + U_g + U_d)^3} \right]^{1/3} = 3 \times \frac{\sqrt[3]{U_n U_g U_d}}{U_n + U_g + U_d} \quad (3)$$

$$T = \alpha U_n + \beta U_g + \delta U_d \quad (4)$$

$$D = (C \times T)^{1/2} \quad (5)$$

where U_n, U_g and U_d denote the levels of new-type urbanization development, greening and digitalization, respectively, and α, β and δ denote the coefficients of the three subsystems. Considering the equal importance of the three subsystems to the overall system, we adopted $\alpha = \beta = \delta = 1/3$ [53,54]; moreover, C has a value ranging from 0 to 1, reflecting the degree of coupling among the NUGD subsystems. In addition, the higher the value is, the lower the degree of dispersion is, i.e., the higher the coupling degree indicates a more refined mechanism of interaction among the three subsystems. T is the comprehensive coordination level of the NUGD system, while D represents the degree of coupling coordination in the NUGD system's development. Moreover, a larger value denotes a greater degree of coordination in the NUGD system. In addition, referring to the studies of Zhang and Chen [55] and Zheng et al. [56], the following classifications were used (Table 2).

Table 2. Classification criteria for CCD.

CCD	Coordination Level	CCD	Coordination Level
[0, 0.1)	Extreme Disorder	[0.5, 0.6)	Barely Coordination
[0.1, 0.2)	Severe Disorder	[0.6, 0.7)	Primary Coordination
[0.2, 0.3)	Moderate Disorder	[0.7, 0.8)	Moderate Coordination
[0.3, 0.4)	Mild Disorder	[0.8, 0.9)	Good Coordination
[0.4, 0.5)	Near Disorder	[0.9, 1.0)	Excellent Coordination

2.3.3. Spatial Association Based on ESDA

To assess whether a region exhibits distinct spatial clustering, we employed the global Moran's I to evaluate the correlation of the collaborative development of the NUGD system among Chinese cities across the entire geographical space. This can be expressed as follows [56]:

$$Moran's\ I = \frac{n}{\sum\limits_i \sum\limits_j W_{ij}} \cdot \frac{\sum\limits_i \sum\limits_j W_{ij}(x_i - \overline{x})(x_j - \overline{x})}{\sum\limits_i (x_i - \overline{x})^2} \tag{6}$$

where the Moran's I range from -1 to 1, with positive (negative) values indicating positive (negative) spatial clustering characteristics. Moreover, its spatial distribution is random, with a value of 0.

The local Moran's I can be used to identify statistically significant spatial outliers. Therefore, the local Moran's I of Anselin (1995) [57] was applied herein to probe into the spatial clustering characteristics of the synergy effect of the NUGD system in Chinese cities. These were characterized using LISA and Moran scatter plots via the following equations:

$$I_i = \frac{x_i - \overline{X}}{S_i^2} \sum_{j=1, j*i}^{n} w_{ij}(x_j - \overline{X}) \tag{7}$$

$$S_i^2 = \frac{\sum\limits_{j=1, j \neq i}^{n} (x_j - \overline{X})^2}{n-1} \tag{8}$$

In the above equations, x_i is the observation value of indicator i, $\overline{X}$ denotes the average value of the corresponding observations and W_{ij} is the spatial weight between elements i and j.

2.3.4. Conditional Kernel Density Estimation

Being a nonparametric estimation method, kernel density estimation can be employed to fit the distribution function according to the characteristics of the data, thus avoiding any errors that may be caused by arbitrarily setting the form of the function. This can effectively address the issue of nonequilibrium. However, the traditional method can only describe the static characteristics of the synergistic effect of the NUGD system and cannot provide a systematic explanation for its dynamic characteristics. To address this issue, conditional kernel density estimation was introduced herein to depict and analyze the distribution features of the collaborative development of NUGD under spatial conditions. The calculation equations for kernel density estimation have been described by Li et al. [58] and Liu et al. [59].

$$f(x) = \frac{1}{Nh} \sum_{i=1}^{N} K\left(\frac{X_i - \overline{x}}{h}\right) \tag{9}$$

$$K(x) = \frac{1}{\sqrt{2\pi}} exp\left(-\frac{x^2}{2}\right) \tag{10}$$

where $f(x)$ represents the density function of the random variable x, $K(\cdot)$ is the Gaussian kernel function, N is the number of observations, X denotes the observed values, $\overline{x}$ denotes the mean of the observed values and h is the bandwidth, revealing the smoothness of the curve and the estimation accuracy. Moreover, a larger bandwidth results in a smoother curve but lower estimation accuracy, and vice versa.

2.3.5. Dagum Gini Coefficient and Its Decomposition

The dynamic variation in the absolute differences in the synergistic NUGD system among the Chinese cities can be described scientifically via kernel density estimation, but this method cannot reveal the trend in the changes in the relative difference. Therefore, it is necessary to rely on the Dagum Gini coefficient to achieve this objective.

In this study, this method was adopted to analyze the regional differences in the triple synergy of the NUGD system, and the differences were decomposed. Moreover, the research objects were divided into k groups, totaling n research objects, where $a, b, \ldots$ represent the different groups; $n_a, n_b, \ldots$ indicate the number of research objects in each group; $y_{ai}, y_{bj}, \ldots$ denote the variable data of any research object within the group that refers to the triple synergy of the NUGD system in the cities; μ is the average synergistic development level; and G denotes the overall Gini coefficient. In addition, the coefficient can be decomposed into the within-region difference contribution (G_w), the between-region contribution (G_{nb}) and the intensity of the transvariation (G_t), and the relationship between these three parts satisfies Equation (12). Generally, a smaller coefficient indicates that there is a closer coupling and coordination level among the cities, and the synergy is strong. The specific calculation process has been described in Dagum [60] and Ji et al. [61].

$$G = \frac{1}{2n^2\mu}\sum_{a=1}^{k}\sum_{b=1}^{k}\sum_{i=1}^{n_a}\sum_{j=1}^{n_b}\left|y_{ai} - y_{bj}\right| \tag{11}$$

$$G = G_w + G_{nb} + G_t \tag{12}$$

2.3.6. Obstacle Degree Model

The multidimensionality of the indicators measuring the collaborative development of the NUGD system determines the complexity of its state and future development. To clarify the obstacles to the synergy of the NUGD and to better understand the reasons for the differences in collaborative development among the different cities, the obstacle degree model was adopted to calculate and rank the target and various indicator layer factors. Then, we analyzed the impact of the various obstacles on the promotion of the collaborative development of the NUGD system in different prefecture-level cities. Finally, we provide helpful proposals to enhance the synergy of the NUGD system. The calculation equation has been introduced in Cheng et al. [62] and encompasses the following:

(1) Calculate the obstacle degree of indicator i in the indicator layer of coupling coordination:

$$O_i = \frac{I_i \times F_i}{\sum\limits_{i=1}^{n} I_i \times F_i} \times 100\% \tag{13}$$

(2) Calculate the obstacle degree of coupling coordination for the criterion layer, denoted as U_j:

$$U_j = \sum o_{ij} \tag{14}$$

where F_i is the factor contribution degree and I_i is the indicator deviation degree.

2.4. Data Sources

Panel data for 282 Chinese cities from 2011 to 2021 were adopted in the study. To thoroughly analyze the internal mechanisms of the synergistic effect among new-type urbanization, greening and digitalization, methods such as the coupling coordination model and ESDA were employed to explore the spatiotemporal variations, agglomeration characteristics and regional differences in the NUGD system. Due to constraints in data integrity, the regions of Hong Kong, Macau and Taiwan, as well as cities with incomplete statistical data or adjusted administrative divisions, such as Bijie, Tongren, Sansha and Hami City, were excluded from this study. The data for the secondary indicators of new-type urbanization, greening and digitalization considered in the system mainly originated from the China City Statistical Yearbook, the China Urban-Rural Construction Statistical Yearbook and the statistical yearbooks of various provinces (cities), while the digital financial inclusion index was obtained from the Digital Finance Research Center of Peking University. In addition, the moving average method was applied to fill in any missing data.

3. Results

3.1. Comprehensive Development Level of NUGD in China

Figure 2 shows the average comprehensive index of the NUGD system and the three subsystems. Overall, the levels of these three subsystems and the corresponding comprehensive level showed relatively consistent upward trends, but the overall level was not high. The comprehensive index steadily increased from 0.109 in 2011 to 0.131 in 2021. Moreover, digital development significantly lagged behind the other two. As for each subsystem, there was a slight decline in digital development from 2011 to 2014; it rebounded in 2015 and then reached its lowest in 2016. Thereafter, it gradually increased until 2021, with the whole showing an upward characteristic. The reason behind this phenomenon may be that the policies and infrastructure related to digital development were not yet of a high standard in the early stages, so they were severely hindered. The new-type urbanization subsystem steadily improved during the sample period, with a particularly notable increase after 2015. The level of the green subsystem was significantly greater than that of the other two subsystems, indicating a gradual upward trend with an increase of 0.042. This occurred because, as ecological and environmental issues become increasingly prominent, green development has been one of China's leading strategies due to its ability to alleviate the conflicts that arise during the process of rapid economic growth and the construction of an ecological civilization.

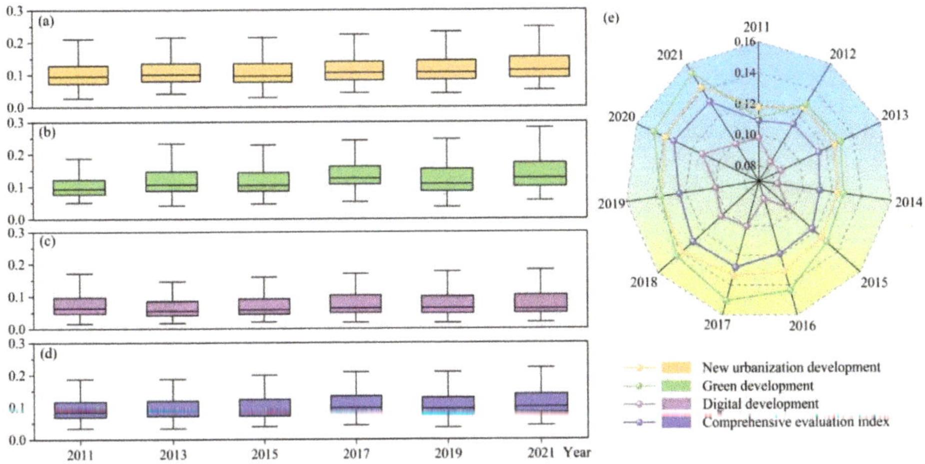

Figure 2. Development of the NUGD subsystems in Chinese cities from 2011 to 2021. (**a**) New urbanization development. (**b**) Green development. (**c**) Digital development. (**d**) Comprehensive evaluation index. (**e**) Development of the three subsystems and NUGD.

3.2. Synergistic Development of NUGD System and Its Spatial Autocorrelation

To accurately determine the temporal variation in the triple synergy of the NUGD system in Chinese cities, the cities under examination were divided into the eastern, central and western regions (Table 3). There was an increase from 0.312 to 0.340 in the average synergy of the NUGD system nationwide, indicating that the synergy of the NUGD system in Chinese cities was continuously strengthened and exhibited steady progress. The average synergistic level of the NUGD system in the three major regions followed the same growth trend as the national average, but there were obvious regional differences between the east and west. In addition, the average synergistic development of the NUGD system in the east was greater than that observed nationwide. The average synergy of the NUGD system in the central and western regions has significantly improved since 2011, with a continuous and steady ascent. Regarding its evolution, the differences between the western and central regions gradually decreased, but there was a gradual acceleration between the eastern and western regions and between the eastern and central regions. This means that the current differences between these two could explain the imbalance at the national level of the NUGD system.

Table 3. Average synergistic level of NUGD nationwide and in three major areas from 2011 to 2021.

Year	Overall	East	Central	West
2011	0.312	0.349	0.295	0.288
2012	0.317	0.353	0.298	0.294
2013	0.318	0.355	0.300	0.294
2014	0.312	0.351	0.293	0.289
2015	0.321	0.363	0.300	0.296
2016	0.320	0.362	0.299	0.296
2017	0.337	0.378	0.316	0.312
2018	0.338	0.381	0.316	0.314
2019	0.331	0.373	0.308	0.307
2020	0.345	0.380	0.325	0.325
2021	0.340	0.380	0.317	0.320

The distribution of the synergistic level in space among the Chinese cities in 2011, 2016 and 2021 is exhibited in Figure 3. There was a stepped distribution pattern from east to west for the NUGD system as a whole. Primary coordination and above were mainly distributed in the east, while synergistic levels were mostly at the moderate and mild disorder levels in the other two regions. This finding indicates that the current synergistic effect of the NUGD system in China is still low, and progress toward a higher coordinated development level is constrained by various factors. Specifically, there were a few cities at the good and moderate coordination levels, and the primary coordination and above occurred mainly in eastern urban clusters and in regions such as the Yangtze River Delta and Pearl River Delta, showing significant spatial agglomeration characteristics. Moreover, the moderate and mild disorder levels were distributed in a continuous urban pattern in the central and western parts, while the borderline disorder and barely coordinated levels were scattered in cities such as Taiyuan, Dalian, Hefei, Zhengzhou and Chongqing. In terms of evolutionary trends, compared to 2011, the number of cities moving from moderate disorder to mild disorder increased in 2021, with a decrease in the fragmented distribution. Moreover, the changes were mainly concentrated in provinces such as Jiangxi, Shaanxi and Gansu, with particularly notable changes in Jiangxi, which may have been due to the radiation-driving effects of the surrounding urban clusters. Additionally, the coordinated development level in various urban clusters remained high, but the only city with good coordination was Shenzhen, showing that the improvement in the coordination level in the cities may have entered a bottleneck period and that local governments must seek a breakthrough and re-evaluate the coordinated development system and mechanism to meet the new requirements proposed and enable coordinated development to enter a new stage.

Moreover, the global Moran's I was obtained using Stata16, after which the local Moran's I was obtained to produce a LISA cluster map to reveal the agglomeration characteristics. As indicated by the results exhibited in Figure 4, the synergistic level of the NUGD system from 2011 to 2021 exhibited significant spatial autocorrelation, with values greater than 0. However, the correlation decreased from 2011 to 2013, gradually increased until 2015 and reached its highest in 2015. Then, from 2015 to 2021, the correlation again decreased. Overall, the global Moran's I of the synergistic level of the NUGD system in Chinese cities slightly decreased, from 0.196 in 2011 to 0.194 in 2021, with an inconspicuous decline, indicating a relatively stable overall spatial agglomeration level.

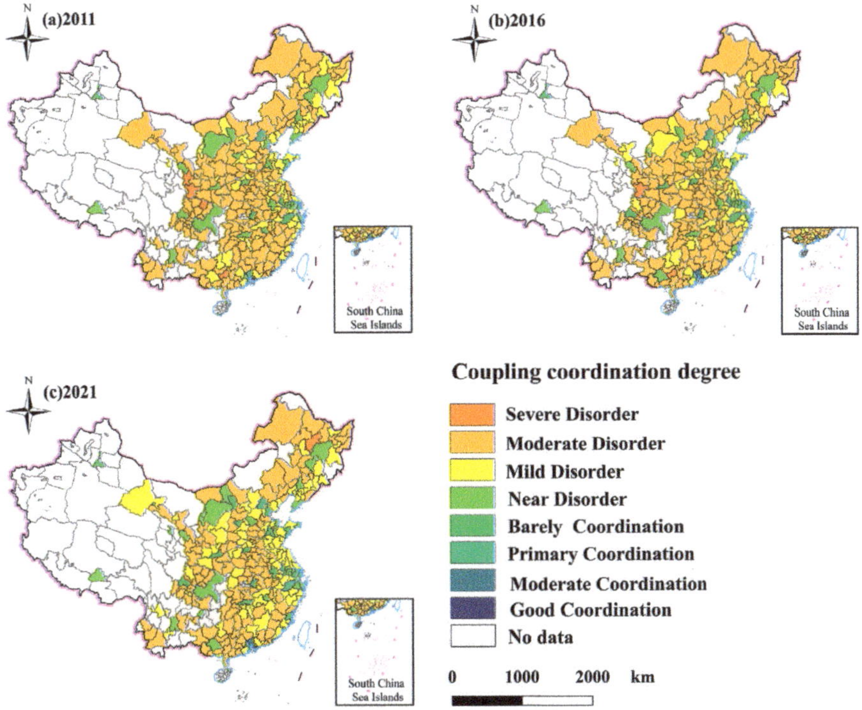

Figure 3. Spatial distribution of synergistic development in 2011, 2016 and 2021. (**a**) 2011. (**b**) 2016. (**c**) 2021.

Figure 4. Global spatial distribution of the NUGD system from 2011 to 2021. (**a**) Moran's *I*. (**b**) Z-Value.

Subsequently, the local Moran's *I* was calculated, and ArcGIS 10.8 and Stata16 were applied to create LISA maps and Moran scatter plots, respectively. The agglomeration patterns of the NUGD system in Chinese cities in 2011, 2016 and 2021 are shown below (Figure 5). Overall, the spatial agglomeration pattern of the synergistic level of the NUGD system in Chinese cities from 2011 to 2021 remained relatively stable, with high–high and low–low as the main types. Specifically, the high–high-type cities were mainly located in the eastern coastal regions, in Shanxi and Jiangxi in the central part and in Inner Mongolia in the western provinces. Furthermore, compared to 2011, there was a growing number of cities reaching a high–low state in 2021, and these cities were mainly located in the west, such as Yunnan, Gansu and Xinjiang. The number and distribution of low–high cities remained relatively stable, and these cities mainly occurred in eastern provinces such as Hebei, Fujian and Shanxi in the central region. The low–low cities were distributed in Yunnan in both 2011 and 2016, but, in 2021, they evolved into high–low areas, with the low–low areas primarily located in Guangxi.

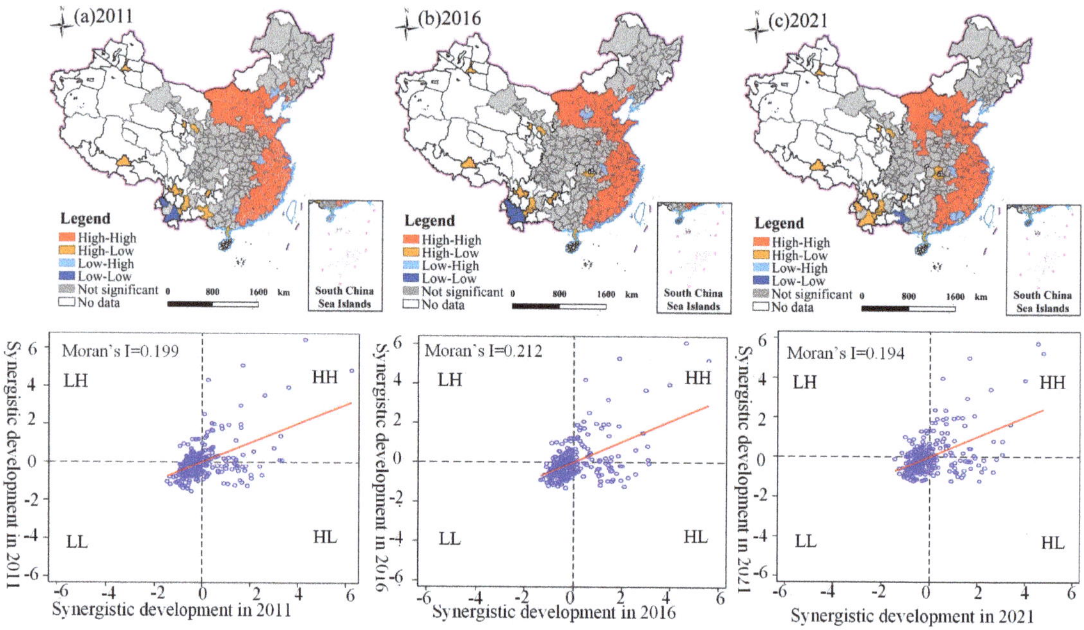

Figure 5. LISA maps and Moran scatter plots for 2011, 2016 and 2021. (**a**) 2011. (**b**) 2016. (**c**) 2021. ***Notes:*** The blue circles represent the individual cities, and the red line represents a linear fit to the scatter.

3.3. Regional Differences in the Synergistic Development of NUGD

3.3.1. Intraregional Differences in the NUGD System

Figure 6a shows that, at the overall level, the intraregional differences were relatively small, showing a W-shaped pattern between 2011 and 2019. Specifically, the national intraregional difference decreased from 2011 to 2012, rebounded and continued to increase until 2015, exhibited a downward trend from 2015 to 2017 and then continued to increase until 2019, reaching the highest of 0.150 during the sample period, after which it slightly decreased. However, the overall national intraregional difference increased during the study period, indicating that there was an increasing trend in the regional imbalance of the NUGD system across China.

Considering the regional differences within each area, synergistic development exhibited the largest differences in the east, followed by the western and central regions. There was fluctuating growth in the eastern and western areas, with trends that could be characterized by inverted W and W shapes, respectively, while the differences within the central region decreased overall over time. In detail, Figure 6a shows that the eastern region experienced a gradual upward trend from 2011 to 2016, a declining trend until 2017, a continued increasing trend until 2019 and a slight decreasing trend thereafter, with an average annual growth rate of 0.42%, which indicates that the regional differences in the east were further expanding. This may be attributed to the high-level economic development effects brought about by the three major urban clusters located in the east, which started the synergistic development of the NUGD system, with continuous and high momentum and a steady improvement in the synergistic process. However, cities in the east, outside the three major urban clusters, are currently attempting to keep pace, but the synergistic level continues to slowly increase, thus widening the gap further.

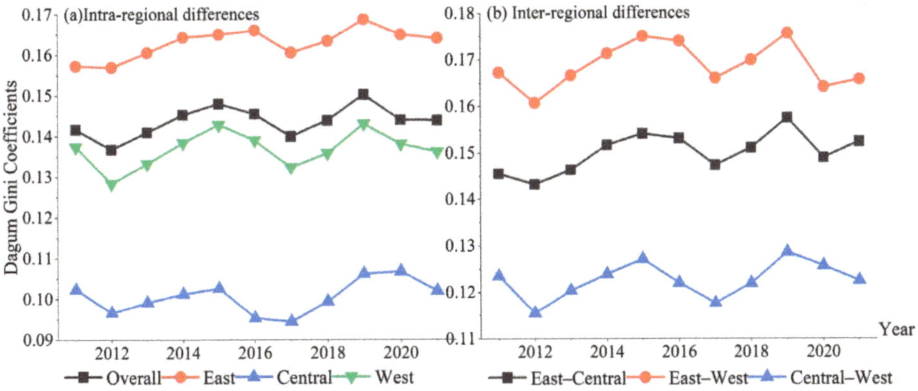

Figure 6. (**a**) Intraregional and (**b**) interregional differences in NUGD system development.

The development trend in the west was consistent with the national synergistic development trend in the NUGD system between 2011 and 2019, with both showing W-shaped patterns. However, there was a significant decline in the western region from 2019 to 2021. The evolutionary trend in the intraregional differences within the central region was similar to that in the west, with regional differences much smaller than those in the western area.

The above reveals that there was an increasing trend in the intraregional differences within the eastern region and the country, while the central and western regions exhibited further decreasing differences. This may have been due to the more extensive implementation of supportive strategies in the western region of China and the rise of Central China, as well as the growing emphasis on the coordinated development of various aspects of the criteria for the evaluation of political performance.

3.3.2. Interregional Differences in the NUGD System

Figure 6b shows that the dynamic evolution trends in the interregional differences could be characterized by a W shape, and all of these trends reached their maximum values within the sample observation window of 2019. Specifically, the interregional differences between the eastern and central regions, as well as between the eastern and western regions, decreased from 2011 to 2012; they then maintained an upward trend until 2015, decreased to 2017 and rebounded in 2019. From 2019 to 2020, the trends again decreased and then increased from 2020 to 2021. However, the differences between the eastern and central regions increased at an average growth rate of 0.464%, while those between the eastern and western regions gradually decreased. However, regarding the numerical values, the difference between the eastern and western regions remained larger than that in the other two, indicating that the national synergistic level of the NUGD system exhibited stepped imbalanced distribution characteristics from east to west. This also indicates that the focus of work to coordinate regional development remained on narrowing the regional gap between the east and west. Throughout the entire examination, the regional imbalance between the central and western parts consistently remained low. Differing from the fluctuations in the eastern–western and eastern–central regional differences from 2019 to 2021, the central–western regional differences showed a downward trend after reaching a peak in 2019, further verifying the policy dividends of the supportive strategy for the western and central areas in China.

3.3.3. Sources and Contribution Rates of Regional Differences

From 2011 to 2021, the contribution rates of the intensity of the transvariation and interregional differences were more volatile than those of the intraregional ones (Table 4). The intensity of the transvariation consistently exceeded that of the interregional and intraregional differences, with average annual values for the intraregional, interregional and transvariation contribution rates of 31.71%, 30.59% and 37.70%, respectively. This indicates that the regional differences in the synergistic NUGD system were mainly caused by the intensity of transvariation. Specifically, the contribution rate of intraregional differences steadily declined until 2016; it then rebounded to 2019, decreased from 2019 to 2020 and subsequently increased again, with the rate slightly increasing. The rate of interregional differences showed a decreasing trend year by year, with an average annual decline rate of 0.69%. The fluctuations during the study period were as follows: a significant decline from 2011 to 2013, a rebound toward 2015, a decrease from 2015 to 2016, a rise from 2016 to 2017 and a continuous decline to 2020, with the minimum value reached during the sample window period. The level finally rebounded from 2020 to 2021. Moreover, the rate of transvariation generally changed in the opposite manner to that of the interregional differences and increased from 36.77% in 2011 to 38.76% in 2021.

Table 4. Sources and contribution rates in the synergistic development of the NUGD system.

Year	Intraregional (G_w)		Interregional (G_{nb})		Intensity of Transvariation (G_t)	
	Contribution Value	Contribution Rate (%)	Contribution Value	Contribution Rate (%)	Contribution Value	Contribution Rate (%)
2011	0.045	31.75	0.045	31.48	0.052	36.77
2012	0.044	31.83	0.042	30.99	0.051	37.18
2013	0.045	31.74	0.044	30.83	0.053	37.43
2014	0.046	31.64	0.045	31.16	0.054	37.20
2015	0.047	31.51	0.047	31.85	0.054	36.64
2016	0.046	31.36	0.047	32.29	0.053	36.35
2017	0.044	31.53	0.044	31.63	0.052	36.84
2018	0.045	31.56	0.045	31.25	0.054	37.19
2019	0.048	31.67	0.046	30.41	0.057	37.92
2020	0.047	32.34	0.036	25.26	0.061	42.40
2021	0.046	31.84	0.042	29.40	0.056	38.76

3.4. *Identification of Obstacles to Synergy in the NUGD System*

Figure 7 presents the average obstacle levels from the criterion and indicator layers for the synergy of the NUGD system in Chinese cities from 2011 to 2021, reflecting the top five obstacles and seven factors influencing synergistic development, respectively. Figure 7a shows that the top five obstacles from 2011 to 2021 were social urbanization (B_3), spatial urbanization (B_4), digital infrastructure (B_8), the digital industry input (B_9) and the digital industry output (B_{10}). Specifically, in 2011, B_3 ranked first, indicating that social urbanization played a restrictive role in the synergy of the NUGD at that time. With the exception of 2011 and 2013, B_4 was always ranked first, indicating that spatial urbanization remained the most important obstacle. B_9 and B_{10} consistently ranked at the lowest two positions from 2011 to 2020, indicating that, during this period, the impacts of the digital industry input and digital industry output on the synergistic development of the NUGD system in the cities were relatively small. However, in 2021, B_9 and B_{10} were the fourth and fifth obstacle factors, respectively, and their obstacle degrees increased, indicating that the obstacle roles of these two factors were enhanced at this time.

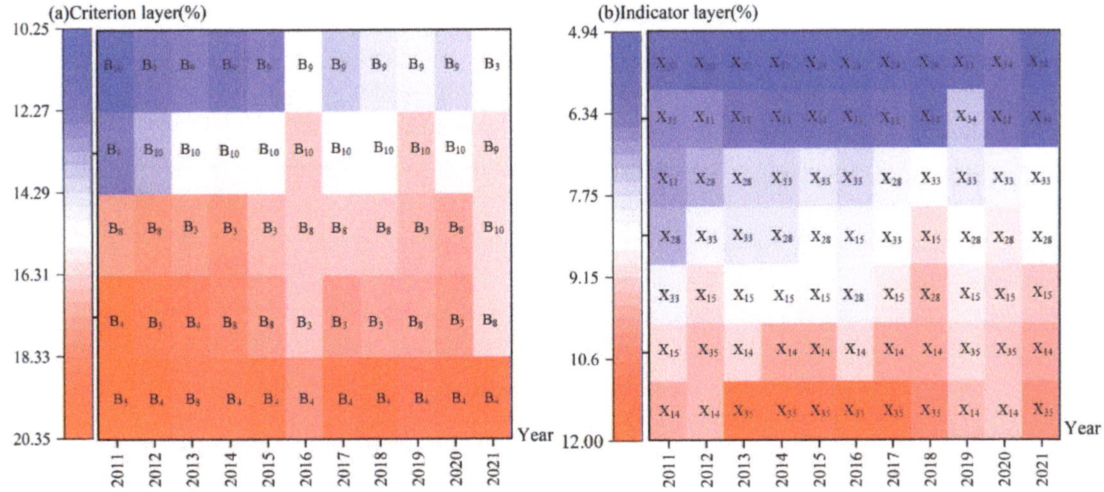

Figure 7. Comparison of the average obstacle degrees of the elements in the criterion (**a**) and indicator layers (**b**) from 2011 to 2021 (%).

The identification of the factors affecting the synergy of the NUGD system based on the criterion layer alone cannot reveal the individual differences between the subindicators. To address this issue, we further calculated the obstacle degree of the indicator layer for synergistic development in Chinese cities from 2011 to 2021 and ranked the main obstacle factors. Figure 7b reflects the top seven main factors in the indicator layer of the synergy of the NUGD system in Chinese cities from 2011 to 2021.

Figure 7b reveals that there were ten factors among the top seven rankings in the indicator layer. First, in the social urbanization dimension, the number of books per 100 people in libraries (X_{11}) and the proportion of college teachers (X_{12}) were major obstacle factors. In terms of their rankings, they were ranked at the bottom and dropped out of the top seven in 2021. Regarding spatial urbanization, the proportion of urban construction land and built-up areas (X_{14} and X_{15}) were major barrier factors. X_{14} ranked first in 2011 and 2012, stabilized at second place from 2013 to 2018, returned to first place in 2019 and 2020 and moved to second place in 2021. The corresponding obstacle level also exhibited fluctuating decline–rise characteristics, identifying X_{14} as a key factor constraining the synergistic development of the NUGD system in Chinese cities. As for X_{15}, its barrier level occurred at an intermediate level, while the magnitude also showed decline–rise characteristics. In the dimension of green living, the number of public transportation vehicles per 10^4 people (X_{20}) was a major obstacle factor from 2011 to 2013, and its obstacle level continuously declined, dropping out of the top seven in the following years, indicating significant progress in urban transportation infrastructure and the transformation toward green and environmental protection. Fourth, considering green ecology, the per capita green park area (X_{24}) was a major obstacle. In terms of its ranking, it ranked seventh from 2015 to 2018 and in 2021, and its obstacle level showed decline–rise evolutionary characteristics. Fifth, in the digital infrastructure dimension, the fiber optic cable density (X_{28}) was a major obstacle factor. In terms of its ranking, the role of X_{28} in constraining the synergistic development of the NUGD system in cities remained at an intermediate level, thereby ranking fourth in most years, ranking fifth in a few years and ranking third in 2015, 2016 and 2018, with the obstacle level further improving. As for the investment in digital industries, the number of higher education students (X_{33}) was a major factor constraining synergistic development in the cities. In terms of its ranking, X_{33} ranked third in 2011, dropped to fourth in 2012 and 2013 and then stabilized at the fifth position, indicating a relatively reduced impact on the synergy of the NUGD system in the cities.

Regarding its numerical value, the obstacle level of X_{33} exhibited fluctuating decline–rise features. The seventh dimension was the output of the digital industry, with the per capita telecommunications business volume (X_{34}) as the main obstacle factor, while the per capita postal business volume (X_{35}) once again became the largest obstacle factor. X_{34} ranked sixth and seventh in 2019 and 2020, respectively, and its obstacle level showed a declining trend, indicating that the internet output of Chinese cities has increased, and that digital development has achieved certain results. Regarding X_{35}, with the exception of 2011, it always occurred within the top two positions, and the obstacle level always remained above 9%.

4. Discussion

4.1. Spatial Disparities in Coupling Coordinated Development across China's Regions

The collaborative development of the NUGD system is vital for Chinese efforts to drive environmentally sound and socially inclusive urbanization practices [63]. By explaining the interaction mechanisms of the three subsystems and referring to existing research, a multielement, multidimensional evaluation indicator was established to uncover the triple synergy of the NUGD system. Moreover, the methods and models used in this study are sound and widely applied in research exploring the coupled coordination between urban ecology and the economy [55,63]. The results reveal that there was continuous improvement and notable agglomeration characteristics in the collaborative development of the NUGD system in China, but the existing regional imbalance was significant. Consistent with the conclusions of most studies on coupling coordination [64], the eastern regions, which benefit from talent resources, policy initiatives and so on, exhibit a greater level of synergy than the other two parts of China. For example, Han et al. (2023) [54] placed the digital economy, technological innovation and ecological environment within the same research framework to examine their temporal and spatial heterogeneity. They found that their coupled coordination level presented the characteristic of "eastern > western" in terms of space. Meanwhile, Ma et al. (2024) [65] considered the coupling coordination level of urbanization and the carbon emission efficiency from a global perspective and systematically analyzed it from three dimensions—the country, developed countries and developing countries—as well as regions, which revealed significant differences in coupling coordination. This is an interesting study that is beneficial for the sustainable development of both individual countries and global cities. Therefore, it is essential to further implement supportive strategies to promote the narrowing of the development gaps between and within regions and countries, ultimately achieving the sustainability of global development.

The analysis of the spatial patterns revealed the static distribution of collaborative development in Chinese cities in multiple dimensions. However, it could not identify the changing features of the triple synergy level in Chinese cities from a dynamic perspective. Therefore, referring to relevant research [59,66], kernel density estimation, including unconditional, spatial static and spatial dynamic kernel density estimation, considering a one-year density, was introduced herein to uncover the distribution dynamics of the synergy of Chinese cities (Figure 8). The unconditional kernel density estimates (Figure 8a) of the probability of the synergistic level of the NUGD system in Chinese cities were distributed along the positive 45° line, which means that, from t to $t + 1$, the synergistic level in Chinese cities has remained relatively stable. However, under static kernel density estimation (Figure 8b), synergistic development revealed significant stage differences in its evolutionary characteristics, showing a fault phenomenon with a boundary at $x = 0.5$. Compared to the spatial static kernel density, the distribution of the dynamic kernel density probability results (Figure 8c) was similar but slightly different, indicating that time indeed impacts the interactions among the different cities regarding synergistic development in China. In conclusion, multidimensional discussions on the spatial disparities in the coordinated development of Chinese urban areas can enrich the related research on new-type urbanization and green and digital development. It also provides new research approaches for cities to reduce regional differences and achieve sustainable and high-quality development.

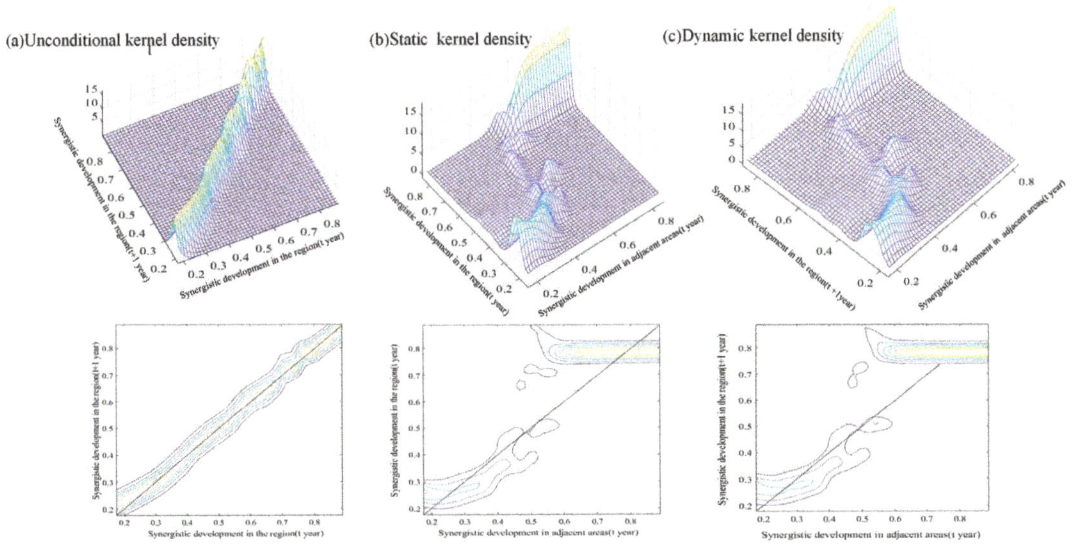

Figure 8. Conditional kernel density and corresponding contour map of NUGD. (**a**) Unconditional kernel density. (**b**) Static kernel density. (**c**) Dynamic kernel density.

Notes: In the kernel density plot, the *X*- and *Y*-axes indicate collaborative development, while the *Z*-axis represents the density (probability) of each point on the *X*–*Y* plane. In the kernel density contour plot, the *X*- and *Y*-axes are the same as those in the kernel density plot. The closer the contour lines are to the center, the greater the value, the denser the contour lines and the greater the change in density, with a convergence trend becoming more evident.

4.2. Major Constraints to the Development of the NUGD System

As previously indicated, regarding the criterion-level factors, the constraint of spatial urbanization (B_4) was the most prominent. Spatial urbanization requires a diversified layout and a coordinated structure, and it can accommodate more of the urban population, promoting urban transportation, culture, entertainment and other municipal and infrastructure construction functions. However, there are still some problems in the process of new-type urbanization, such as the disorderly expansion of urban land and the imbalanced development of the spatial distribution [67]. Moreover, the ratio of urban construction land's occupation to its utilization efficiency has been shown to be negative, further exacerbating the severe situation of urban resource constraints, increased pollution and decreased urban ecological resilience, thus constraining the comprehensive improvement of the synergy of the NUGD system in cities [68]. Moreover, as the city expands and the population increases, the pressure of constructing and maintaining urban infrastructure also grows. In some cases, the pace of urban digital infrastructure development may not reflect the speed of urban expansion, thus affecting the progress of digitalization. Therefore, at present, it is crucial to accelerate the optimization of the spatial layout and structure, improve the efficiency of spatial allocation and consistently implement major regional collaborative development strategies to adapt to the spatial needs for the high-level development of the NUGD system.

In regard to the constraints at the index layer, the proportion of urban construction land (X_{14}) and per capita postal business volume (X_{35}) were major constraints. Due to the Chinese reform and opening-up process, the urban land area has expanded significantly, leading to an increase in the urban heat island effect and global warming, degrading natural landscapes and changing land use patterns [68–70], which in turn has hindered further improvements in the synergy of the NUGD system in cities. The per capita postal business volume (X_{35}) once became the largest obstacle factor. The reasons for this may be as follows: due to the increasingly prevalent e-commerce, the per capita volume of postal services has increased, while the demand for personnel for logistics services has also increased, failure to optimize logistics and transportation modes may lead to wasted resources and increased CO2 emissions in the logistics process [71]. Moreover, the resource utilization of the existing logistics industry is becoming saturated. Therefore, in addition to providing traditional infrastructure support and hardware and software resources, it is necessary to invest large amounts of funds in new infrastructure construction to promote the digital transformation, expand the logistics resources and optimize resource allocation to finally enhance the supply chain's resilience for the high-level synergy of the NUGD [72,73].

4.3. Limitations and Prospects

In this study of the coupling coordination level of new-type urbanization, greening and digitalization, we have thoroughly demonstrated why and how these three subsystems exhibit synergy. We employed various methods and presented their spatiotemporal differentiation and regional differences in a visual graph format. Finally, we used the obstacle degree model to reveal the main factors affecting the development of this system at the criterion and indicator levels. This is of great significance in enabling China to achieve new progress in economic development, social welfare and the construction of an ecological civilization. In summary, it is essential for the sustainable development of cities. However, the limitations specified below exist.

Firstly, due to the limited knowledge of the levels of urbanization, greening and digitalization in other countries around the world, as well as difficulties in obtaining relevant data, this study mainly took China as an example, which is a shortcoming in the scope of this research. However, we sincerely hope that the research can serve as a reference for other countries and regions facing the adverse impacts of rapid urbanization and eager to achieve sustainable urban development through green transformation and digital empowerment. Secondly, compared to previous studies, our focus was on the interrelationship between new-type urbanization, greening and digitalization. This study did not explore the possible causal relationships among these three areas. Furthermore, our research was not extended to other fields—for example, whether urbanization has increased the mortality rate of birds; whether it has caused changes in reptile communities; and why these phenomena occur. These are all highly interesting research topics that are of great significance to the sustainable development of cities. In the future, we will examine urbanization, green development and digital transformation from a global perspective, extend our research to other fields, further explore these three areas and compare the situation in other countries with that in China in order to draw more universal conclusions.

5. Conclusions and Policy Implications

Based on the realistic goals of improving economic development efficiency, coordinating rural revitalization, promoting green development and fostering harmony between humans and nature, data from 2011 to 2021 for 282 cities in China were adopted herein as a research sample to scientifically uncover the triple synergy effect of the NUGD system in Chinese cities. The basic conclusions of this study are listed below.

First, regarding the comprehensive level of the NUGD system in Chinese cities, the levels of new-type urbanization and green and comprehensive development showed consistent upward trends, while that of digitalization lagged significantly behind that of the other two subsystems, with moderate overall development levels.

Second, regarding the triple synergy effect of the NUGD system in Chinese cities, over time, there has been an upward trend both at the national and regional level, but the overall development quality is not high. In terms of spatial patterns, significant spatial differentiation was observed, with high values in the east and low values in the west. Specifically, cities with primary and higher levels of coordinated development were located in regions and cities with high development levels, while many central and western cities exhibited a moderate disorder level. Considering spatial agglomeration, the agglomeration level throughout the whole NUGD system remained relatively stable, mainly characterized by the "high–high" and "low–low" types.

Third, regarding regional differences, there were increasing trends at the national and the eastern level, while those in the other two parts were narrowing further. Moreover, the difference between the eastern and western regions was always larger than that between the eastern and central regions and that between the central and western regions, highlighting the need to accelerate their coordinated regional development. Notably, the regional differences in the synergistic development of the NUGD system were mainly caused by the intensity of the transvariation.

Fourth, the obstacle factor analysis revealed that social urbanization (B_3), spatial urbanization (B_4), digital infrastructure (B_8), the digital industry input (B_9) and the digital industry output (B_{10}) were the main obstacle factors at the criterion level, while the proportion of urban construction land and built-up areas (X_{14} and X_{15}) and the per capita postal business volume (X_{35}) were the main obstacle factors at the indicator level.

Policy Implications

For the improvement of the synergy effect of the NUGD system in China, and in order to achieve steady and balanced progress, the following four policy recommendations can be formulated (Figure 9).

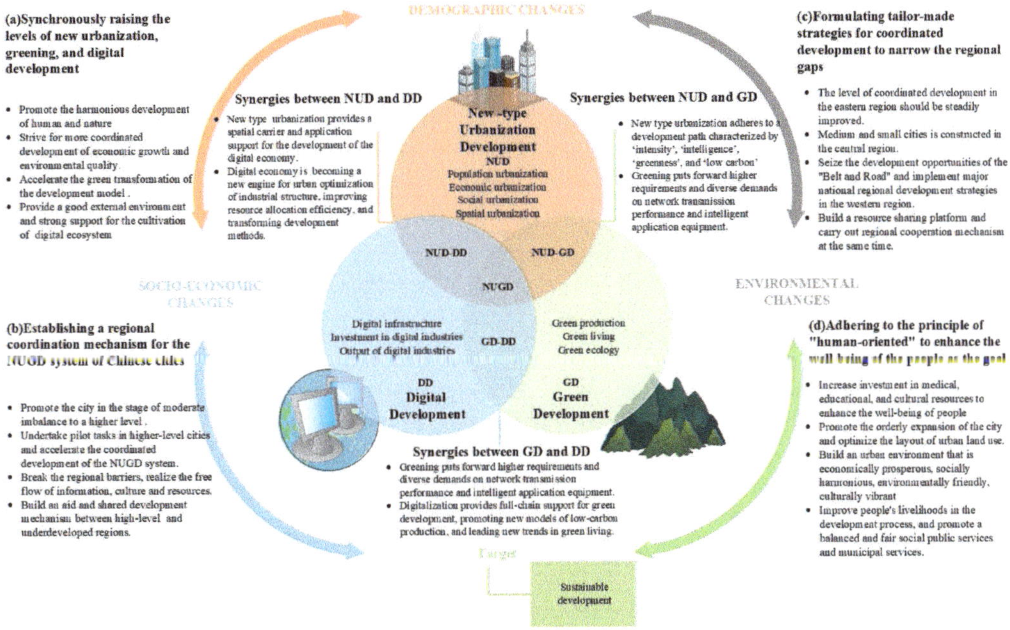

Figure 9. Policy recommendations.

First, efforts should be focused on the synchronous promotion of new-type urbanization, greening and digital development in Chinese cities in order to improve their overall levels and the quality of the synergy and ultimately realize the goals of substantial and sustainable development (Figure 9a). Guided by the five major development concepts, local governments should focus on adhering to urbanization with a human-centered approach, ensuring strong progress in economic growth and eco-quality. Meanwhile, governments should provide strong support to accelerate the green transformation of development patterns, promoting the formation of environmentally friendly production methods and lifestyles. In addition, they must continuously strengthen the collaborative prevention and control of pollutants, as well as improve the ecological protection compensation system and its levels, thereby enhancing the diversity and stability of ecosystems. Moreover, positive experiences and practices should be actively shared with other countries. Finally, it is essential to provide a favorable external environment and strong support to cultivate a robust digital ecosystem, vigorously promote the industrialization of digital technology and the digitalization of industries and further deepen the integration between the internet and the real economy.

Second, to adhere to a unified national strategy, we should establish a regional coordination mechanism for the NUGD system in Chinese cities (Figure 9b). There were significant agglomeration characteristics in the NUGD system in the Chinese cities studied, and the synergy effect of the NUGD system in neighboring cities significantly impacted the local dynamic evolution. Therefore, we must fully leverage the spatial linkage role of the synergistic development of the NUGD system. Specifically, cities or regions with a higher coordinated development level for the NUGD system should actively take on pilot tasks and accumulate experience and lessons to promote synergistic NUGD system development, providing reference value for cities or regions with lagging levels so as to accelerate synergistic system development. Similarly, countries should enhance cooperation and exchange among themselves, led by developed countries and enabling less developed countries to keep pace, ultimately achieving the sustainable development of the global economic, social and ecological systems. In addition, governments should actively eliminate regional barriers to achieve the free flow of information, culture, resources, etc., between cities, regions and countries. They should promote interregional exchange and cooperation, form a mutually reinforcing complementary system and further construct a mechanism for assistance and shared development between regions with a high synergistic development level for the NUGD system and underdeveloped regions, promoting cities to transition from a moderately imbalanced phase to a higher coordinated development stage.

Third, tailor-made strategies for coordinated development should be formulated to mitigate the regional imbalances in the synergistic NUGD system (Figure 9c). The eastern region should fully leverage its advantages of higher economic development levels, talent aggregation and advanced technology and steadily enhance the synergistic development level of the NUGD system by cultivating interdisciplinary talent and developing high-tech industries. The central region should rely closely on urban clusters and metropolitan areas; eliminate regional barriers; promote the flow of resources, technology and talent; and strengthen its interregional communication and cooperation, thereby constructing a development pattern that encompasses large, medium and small cities to further improve the level and quality of the synergy in the NUGD system. It is vital for the western region to seize its development opportunities, especially the Belt and Road Initiative; implement major national regional development strategies; actively construct a regional cooperation mechanism for resource sharing, platform co-building and talent sharing; and proactively learn from the experiences and lessons of advanced regions regarding the synergistic development of the NUGD system.

Fourth, local governments must take human-centeredness as a premise and increase their investment in medical, educational and cultural resources to enhance the well-being of their citizens (Figure 9d). More importantly, with spatial urbanization, especially urban construction land, becoming the largest obstacle factor, the local government should strive to promote the orderly and rational expansion of cities and optimize the layout of urban land use. This means that, within urban land use, they should control the proportion of production land, ensure living land and increase ecological land. By adjusting the internal structure and layout of urban land use, local governments will be able to build an urban environment that is economically prosperous, socially harmonious, resource-efficient, environmentally friendly, culturally vibrant and ecologically livable. Lastly, the government should ensure and improve the livelihoods of their citizens in the development process, enable them to fully enjoy the benefits brought by social, public and municipal services, and finally enhance their well-being and improve their quality of life.

Author Contributions: Conceptualization, Q.L.; methodology, Q.L., J.G. and H.F.; software, J.G. and M.H.; validation, Q.L. and J.G.; formal analysis, J.G. and H.F.; investigation, J.G. and X.W.; resources, Q.L. and J.G.; data curation, Q.L., J.G. and H.F.; writing—original draft preparation, J.G., X.W. and M.H.; writing—review and editing, J.G., M.H. and H.F.; visualization, J.G. and H.F.; supervision, Q.L.; project administration Q.L. and H.F.; funding acquisition, Q.L. and J.G. All authors have read and agreed to the published version of the manuscript.

Funding: This research was funded by Q.L.'s Guangxi Science and Technology Base and Talent Special Project (Grant No. 2022AC21262), the Planning Research Project of Guangxi Philosophy and Social Science (Grant No. 21CYJ016) and J.G.'s Innovation Project of the School of Economics and Management, Guangxi Normal University (Grant No. JGYJS2024002).

Data Availability Statement: The raw data supporting the conclusions of this article will be made available by the authors on request.

Acknowledgments: We are so grateful to the anonymous reviewers and editors for their suggestions.

Conflicts of Interest: The authors declare no conflicts of interest.

References

1. Jiang, Z.; Zhang, X.; Zhao, Y.; Li, C.; Wang, Z. The Impact of Urban Digital Transformation on Resource Sustainability: Evidence from a Quasi-Natural Experiment in China. *Resour. Policy* **2023**, *85*, 103784. [CrossRef]
2. Zhu, C.; Fang, C.; Zhang, L. Analysis of the Coupling Coordinated Development of the Population–Water–Ecology–Economy System in Urban Agglomerations and Obstacle Factors Discrimination: A Case Study of the Tianshan North Slope Urban Agglomeration, China. *Sustain. Cities Soc.* **2023**, *90*, 104359. [CrossRef]
3. Adam, D. How far will global population rise? Researchers can't agree. *Nature* **2021**, *597*, 462–465. [CrossRef] [PubMed]
4. Inostroza, L.; Taubenböck, H. Searching for the DNA of Urbanisation. A Material Perspective. *Cities* **2024**, *151*, 105079. [CrossRef]
5. Wu, Y.; Zong, T.; Shuai, C.; Jiao, L. How Does New-Type Urbanization Affect Total Carbon Emissions, per Capita Carbon Emissions, and Carbon Emission Intensity? An Empirical Analysis of the Yangtze River Economic Belt, China. *J. Environ. Manag.* **2024**, *349*, 119441. [CrossRef]
6. Tripathi, V.; Edrisi, S.A.; Chen, B.; Gupta, V.K.; Vilu, R.; Gathergood, N.; Abhilash, P.C. Biotechnological Advances for Restoring Degraded Land for Sustainable Development. *Trends Biotechnol.* **2017**, *35*, 847–859. [CrossRef]
7. Jia, W.; Zhao, S.; Liu, S. Vegetation Growth Enhancement in Urban Environments of the Conterminous United States. *Glob. Chang. Biol.* **2018**, *24*, 4084–4094. [CrossRef]
8. Lanka, A.; Dimante-Deimantovica, I.; Saarni, S.; Stivrins, N.; Tylmann, W.; Zawiska, I.; Veski, S. Urbanization-driven Cladocera community shifts in the lake-a case study from Baltic region, Europe. *Anthropocene* **2024**, *46*, 100439. [CrossRef]
9. Whitney, J.W.; Glancy, P.A.; Buckingham, S.E.; Ehrenberg, A.C. Effects of Rapid Urbanization on Streamflow, Erosion, and Sedimentation in a Desert Stream in the American Southwest. *Anthropocene* **2015**, *10*, 29–42. [CrossRef]
10. Tratalos, J.; Fuller, R.A.; Warren, P.H.; Davies, R.G.; Gaston, K.J. Urban form, biodiversity potential and ecosystem services. *Landsc. Urban Plan.* **2007**, *83*, 308–317. [CrossRef]
11. Wu, H.; Hao, Y.; Weng, J.-H. How Does Energy Consumption Affect China's Urbanization? New Evidence from Dynamic Threshold Panel Models. *Energ Policy* **2019**, *127*, 24–38. [CrossRef]
12. Cai, J.; Li, X.; Liu, L.; Chen, Y.; Wang, X.; Lu, S. Coupling and Coordinated Development of New Urbanization and Agro-Ecological Environment in China. *Sci. Total Environ.* **2021**, *776*, 145837. [CrossRef] [PubMed]
13. Buzási, A.; Jäger, B.S. District-Scale Assessment of Urban Sustainability. *Sustain.Cities Soc.* **2020**, *62*, 102388. [CrossRef]
14. Zeng, L.; Wang, J.; Zhang, J.; Lv, J.; Cui, W. New Urbanization Paths in Mineral Resource Abundant Regions in China: A Three-Dimensional Cube Framework. *Resour. Policy* **2020**, *68*, 101709. [CrossRef]
15. Cheng, Z.; Li, X.; Zhang, Q. Can New-Type Urbanization Promote the Green Intensive Use of Land? *J. Environ. Manag.* **2023**, *342*, 118150. [CrossRef] [PubMed]
16. Zhang, J.; Zhang, P.; Wang, R.; Liu, Y.; Lu, S. Identifying the Coupling Coordination Relationship between Urbanization and Forest Ecological Security and Its Impact Mechanism: Case Study of the Yangtze River Economic Belt, China. *J. Environ. Manag.* **2023**, *342*, 118327. [CrossRef] [PubMed]
17. Liu, K.; Mahmoud, H.A.; Liu, L.; Halteh, K.; Arnone, G.; Shukurullaevich, N.K.; Alzoubi, H.M. Exploring the Nexus between Fintech, Natural Resources, Urbanization, and Environment Sustainability in China: A QARDL Study. *Resour. Policy* **2024**, *89*, 104557. [CrossRef]
18. Xu, L.; Fan, M.; Yang, L.; Shao, S. Heterogeneous Green Innovations and Carbon Emission Performance: Evidence at China's City Level. *Energ Econ.* **2021**, *99*, 105269. [CrossRef]

19. Yu, S.; Wang, X.; Liu, J.; Wei, F. Role of Mining Waste Trade on Green Development in China: Policy Implications for Circular Economy. *Resour. Policy* **2023**, *86*, 104147. [CrossRef]
20. Wang, M.; Feng, C. Revealing the Pattern and Evolution of Global Green Development between Different Income Groups: A Global Meta-Frontier by-Production Technology Approach. *Environ. Impact Assess. Rev.* **2021**, *89*, 106600. [CrossRef]
21. Jiang, L.; He, S.; Zhou, H.; Kong, H.; Wang, J.; Cui, Y.; Wang, L. Coordination between Sulfur Dioxide Pollution Control and Rapid Economic Growth in China: Evidence from Satellite Observations and Spatial Econometric Models. *Struct. Change Econ. Dynam.* **2021**, *57*, 279–291. [CrossRef]
22. Liu, H.; Liu, Z.; Zhang, C.; Li, T. Transformational Insurance and Green Credit Incentive Policies as Financial Mechanisms for Green Energy Transitions and Low-Carbon Economic Development. *Energ Econ.* **2023**, *126*, 107016. [CrossRef]
23. Li, J.; Yuan, W.; Qin, X.; Qi, X.; Meng, L. Coupling Coordination Degree for Urban Green Growth between Public Demand and Government Supply in Urban Agglomeration: A Case Study from China. *J. Environ. Manag.* **2022**, *304*, 114209. [CrossRef] [PubMed]
24. Ling, S.; Jin, S.; Wang, H.; Zhang, Z.; Feng, Y. Transportation Infrastructure Upgrading and Green Development Efficiency: Empirical Analysis with Double Machine Learning Method. *J. Environ. Manag.* **2024**, *358*, 120922. [CrossRef] [PubMed]
25. Zeng, S.; Gu, J. Coordination Evaluation and Dynamic Adjustment Mechanism of China's Green Development at Inter-Provincial Level. *Ecol. Indic.* **2023**, *153*, 110419. [CrossRef]
26. Wang, L.; Chen, Y.; Ramsey, T.S.; Hewings, G.J.D. Will Researching Digital Technology Really Empower Green Development? *Technol. Soc.* **2021**, *66*, 101638. [CrossRef]
27. Ma, D.; Zhu, Q. Innovation in Emerging Economies: Research on the Digital Economy Driving High-Quality Green Development. *J. Bus. Res.* **2022**, *145*, 801–813. [CrossRef]
28. Liu, X.; Liu, F.; Ren, X. Firms' Digitalization in Manufacturing and the Structure and Direction of Green Innovation. *J. Environ. Manag.* **2023**, *335*, 117525. [CrossRef] [PubMed]
29. Grybauskas, A.; Stefanini, A.; Ghobakhloo, M. Social Sustainability in the Age of Digitalization: A Systematic Literature Review on the Social Implications of Industry 4.0. *Technol. Soc.* **2022**, *70*, 101997. [CrossRef]
30. Brenner, B.; Hartl, B. The Perceived Relationship between Digitalization and Ecological, Economic, and Social Sustainability. *J. Clean. Prod.* **2021**, *315*, 128128. [CrossRef]
31. Wang, X.; Qin, C.; Liu, Y.; Tanasescu, C.; Bao, J. Emerging Enablers of Green Low-Carbon Development: Do Digital Economy and Open Innovation Matter? *Energ Econ.* **2023**, *127*, 107065. [CrossRef]
32. Tang, M.; Liu, Y.; Hu, F.; Wu, B. Effect of Digital Transformation on Enterprises' Green Innovation: Empirical Evidence from Listed Companies in China. *Energ Econ.* **2023**, *128*, 107135. [CrossRef]
33. Liang, C.; Chen, X.; Di, Q. Path to Pollution and Carbon Reduction Synergy from the Perspective of the Digital Economy: Fresh Evidence from 292 Prefecture-Level Cities in China. *Environ. Res.* **2024**, *252*, 119050. [CrossRef] [PubMed]
34. Fang, C.; Cui, X.; Li, G.; Bao, C.; Wang, Z.; Ma, H.; Sun, S.; Liu, H.; Luo, K.; Ren, Y. Modeling Regional Sustainable Development Scenarios Using the Urbanization and Eco-Environment Coupler: Case Study of Beijing-Tianjin-Hebei Urban Agglomeration, China. *Sci. Total Environ.* **2019**, *689*, 820–830. [CrossRef] [PubMed]
35. Song, S.; Wen, J.; Li, Y.; Li, L. How Does Digital Economy Affect Green Technological Innovation in China? New Evidence from the "Broadband China" Policy. *Econ. Anal. Pol.* **2024**, *81*, 1093–1112. [CrossRef]
36. Lin, B.; Zhu, J. Impact of China's New-Type Urbanization on Energy Intensity: A City-Level Analysis. *Energy Econ.* **2021**, *99*, 105292. [CrossRef]
37. Lv, L.; Chen, Y. The Collision of Digital and Green: Digital Transformation and Green Economic Efficiency. *J. Environ. Manag.* **2024**, *351*, 119906. [CrossRef] [PubMed]
38. Wang, D.; Zhou, T.; Wang, M. Information and Communication Technology (ICT), Digital Divide and Urbanization: Evidence from Chinese Cities. *Technol. Soc.* **2021**, *64*, 101516. [CrossRef]
39. Zhao, Y.; Song, Z.; Chen, J.; Dai, W. The Mediating Effect of Urbanisation on Digital Technology Policy and Economic Development: Evidence from China. *J. Innov. Knowl.* **2023**, *8*, 100318. [CrossRef]
40. Ma, Q.; Shi, F. New Urbanization and High-Quality Urban and Rural Development: Based on the Interactive Coupling Analysis of Industrial Green Transformation. *Ecol. Indic.* **2023**, *156*, 111044. [CrossRef]
41. Yu, B. Ecological Effects of New-Type Urbanization in China. *Renew. Sustain. Energy Rev.* **2021**, *135*, 110239. [CrossRef]
42. Artmann, M.; Kohler, M.; Meinel, G.; Gan, J.; Ioja, I.-C. How Smart Growth and Green Infrastructure Can Mutually Support Each Other—A Conceptual Framework for Compact and Green Cities. *Ecol. Indic.* **2019**, *96*, 10–22. [CrossRef]
43. Zhang, F.; Chung, C.K.L.; Yin, Z. Green Infrastructure for China's New Urbanisation: A Case Study of Greenway Development in Maanshan. *Urban Stud.* **2020**, *57*, 508–524. [CrossRef]
44. Shang, J.; Wang, Z.; Li, L.; Chen, Y.; Li, P. A Study on the Correlation between Technology Innovation and the New-Type Urbanization in Shaanxi Province. *Technol. Forecast. Soc.* **2018**, *135*, 266–273. [CrossRef]
45. Ahirwar, R.; Tripathi, A.K. E-Waste Management: A Review of Recycling Process, Environmental and Occupational Health Hazards, and Potential Solutions. *Environ. Nanotechnol. Monit. Manag.* **2021**, *15*, 100409. [CrossRef]
46. Guo, J.; Wang, L.; Zhou, W.; Wei, C. Powering Green Digitalization: Evidence from 5G Network Infrastructure in China. *Resour. Conserv. Recy.* **2022**, *182*, 106286. [CrossRef]

47. Pan, W.; Xie, T.; Wang, Z.; Ma, L. Digital Economy: An Innovation Driver for Total Factor Productivity. *J. Bus. Res.* **2022**, *139*, 303–311. [CrossRef]
48. Dian, J.; Song, T.; Li, S. Facilitating or Inhibiting? Spatial Effects of the Digital Economy Affecting Urban Green Technology Innovation. *Energy Econ.* **2024**, *129*, 107223. [CrossRef]
49. He, B.; Du, X.; Lu, Y.; Chen, Q.; Lan, R. An Improved Approach for Measuring the Coupling Relationship between New Type Urbanization and Low Carbon Development in China. *Ecol. Indic.* **2024**, *158*, 111383. [CrossRef]
50. Tian, Y.; Wang, R.; Liu, L.; Ren, Y. A Spatial Effect Study on Financial Agglomeration Promoting the Green Development of Urban Agglomerations. *Sustain. Cities Soc.* **2021**, *70*, 102900. [CrossRef]
51. Gao, D.; Li, G.; Yu, J. Does Digitization Improve Green Total Factor Energy Efficiency? Evidence from Chinese 213 Cities. *Energy* **2022**, *247*, 123395. [CrossRef]
52. Yi, M.; Liu, Y.; Sheng, M.S.; Wen, L. Effects of Digital Economy on Carbon Emission Reduction: New Evidence from China. *Energy Policy* **2022**, *171*, 113271. [CrossRef]
53. Dong, L.; Longwu, L.; Zhenbo, W.; Liangkan, C.; Faming, Z. Exploration of Coupling Effects in the Economy–Society–Environment System in Urban Areas: Case Study of the Yangtze River Delta Urban Agglomeration. *Ecol. Indic.* **2021**, *128*, 107858. [CrossRef]
54. Han, X.; Fu, L.; Lv, C.; Peng, J. Measurement and Spatio-Temporal Heterogeneity Analysis of the Coupling Coordinated Development among the Digital Economy, Technological Innovation and Ecological Environment. *Ecol. Indic.* **2023**, *151*, 110325. [CrossRef]
55. Zhang, D.; Chen, Y. Evaluation on Urban Environmental Sustainability and Coupling Coordination among Its Dimensions: A Case Study of Shandong Province, China. *Sustain. Cities Soc.* **2021**, *75*, 103351. [CrossRef]
56. Zheng, H.; Khan, Y.A.; Abbas, S.Z. Exploration on the Coordinated Development of Urbanization and the Eco-Environmental System in Central China. *Environ. Res.* **2022**, *204*, 112097. [CrossRef] [PubMed]
57. Anselin, L. Local indicators of spatial association—LISA. *Geogr. Anal.* **1995**, *27*, 93–115. [CrossRef]
58. Li, S.; Zhang, H.; Li, W. Regional differences and dynamic evolution of China's inclusive finance. *J. Quant. Tech. Econ.* **2019**, *36*, 62–80. (In Chinese)
59. Liu, H.; Qiao, L.; Sun, S. Spatial distribution and dynamic change of water use efficiency in the Yellow River Basin. *Res. Sci.* **2020**, *42*, 57–68. (In Chinese) [CrossRef]
60. Dagum, C. A New Approach to the Decomposition of the Gini Income Inequality Ratio. *Empir. Econ.* **1997**, *22*, 515–531. [CrossRef]
61. Ji, J.; Liu, H.; Yin, X. Evaluation and Regional Differences Analysis of the Marine Industry Development Level: The Case of China. *Mar. Policy* **2023**, *148*, 105445. [CrossRef]
62. Cheng, H.; Zhu, L.; Meng, J. Fuzzy Evaluation of the Ecological Security of Land Resources in Mainland China Based on the Pressure-State-Response Framework. *Sci. Total Environ.* **2022**, *804*, 150053. [CrossRef] [PubMed]
63. Zhang, X.; Fan, H.; Liu, F.; Lv, T.; Sun, L.; Li, Z.; Shang, W.; Xu, G. Coupling Coordination between the Ecological Environment and Urbanization in the Middle Reaches of the Yangtze River Urban Agglomeration. *Urban Clim.* **2023**, *52*, 101698. [CrossRef]
64. Ge, Y.; Hu, S.; Song, Y.; Zheng, H.; Liu, Y.; Ye, X.; Ma, T.; Liu, M.; Zhou, C. Sustainable Poverty Reduction Models for the Coordinated Development of the Social Economy and Environment in China. *Sci. Bull.* **2023**, *68*, 2236–2246. [CrossRef]
65. Ma, D.; Yan, Y.; Xiao, Y.; Zhang, F.; Zha, H.; Chang, R.; Zhang, J.; Guo, Z.; An, B. Research on the Spatiotemporal Evolution and Influencing Factors of Urbanization and Carbon Emission Efficiency Coupling Coordination: From the Perspective of Global Countries. *J. Environ. Manag.* **2024**, *360*, 121153. [CrossRef]
66. Lee, J.; Gong, J.; Li, S. Exploring spatiotemporal clusters based on extended kernel estimation methods. *Int. J. Geogr. Inf. Sci.* **2017**, *31*, 1154–1177.
67. Liu, Y.; Zhang, X.; Kong, X.; Wang, R.; Chen, L. Identifying the Relationship between Urban Land Expansion and Human Activities in the Yangtze River Economic Belt, China. *Appl. Geogr.* **2018**, *94*, 163–177. [CrossRef]
68. Feng, X.; Xiu, C.; Bai, L.; Zhong, Y.; Wei, Y. Comprehensive Evaluation of Urban Resilience Based on the Perspective of Landscape Pattern: A Case Study of Shenyang City. *Cities* **2020**, *104*, 102722. [CrossRef]
69. Chen, M.; Liu, W.; Lu, D. Challenges and the Way Forward in China's New-Type Urbanization. *Land Use Policy* **2016**, *55*, 334–339. [CrossRef]
70. Hou, D.; Meng, F.; Prishchepov, A.V. How Is Urbanization Shaping Agricultural Land-Use? Unraveling the Nexus between Farmland Abandonment and Urbanization in China. *Landsc. Urban Plan.* **2021**, *214*, 104170. [CrossRef]
71. Kaur, H.; Singh, S.P. Modeling Low Carbon Procurement and Logistics in Supply Chain: A Key towards Sustainable Production. *Sustain. Prod. Consum.* **2017**, *11*, 5–17. [CrossRef]
72. Xu, A.; Qian, F.; Ding, H.; Zhang, X. Digitalization of Logistics for Transition to a Resource-Efficient and Circular Economy. *Resour. Policy* **2023**, *83*, 103616. [CrossRef]
73. Jayarathna, C.P.; Agdas, D.; Dawes, L. Exploring Sustainable Logistics Practices toward a Circular Economy: A Value Creation Perspective. *Bus. Strateg. Environ.* **2023**, *32*, 704–720. [CrossRef]

land

MDPI

Article

Effect of Artificial Intelligence on Chinese Urban Green Total Factor Productivity

Yuanhe Zhang [1,2] and Chaobo Zhou [3,4,*]

[1] Dong Fureng Institute of Economic and Social Development, Wuhan University, Wuhan 430072, China
[2] Economics and Management School, Wuhan University, Wuhan 430072, China
[3] College of International Economics and Trade, Ningbo University of Finance and Economics, Ningbo 315175, China
[4] Climate Change and Energy Economics Study Center of Wuhan University, Wuhan 430072, China
* Correspondence: zhouchaobo@whu.edu.cn

Abstract: The manner of achieving high-quality economic development in China through artificial intelligence (AI) has become a focus of academic attention. On the basis of panel data of prefecture-level cities in China from 2010 to 2021, this research utilizes the exogenous impact of the implementation of the National New Generation Artificial Intelligence Innovation and Development Pilot Zone (AIPZ) to explore the causal effect between AI and green total factor productivity (GTFP). The results are as follows: (1) AI has a significant enhancement effect on urban GTFP. After using a series of robustness tests, such as parallel trend sensitivity test, heterogeneity treatment effect test, and machine learning, this conclusion remains robust. (2) Subsequent mechanism analysis shows that the impact of AI on urban GTFP is mainly achieved by enhancing urban green innovation, promoting industrial structure upgrading, and reducing land resource misallocation. (3) Lastly, the effect of AI on urban GTFP is heterogeneous. AI has also markedly significant enhancement effects on high human capital, non-resource-based economies, and high levels of green consumption behavior. This study provides useful insights for China to develop AI and achieve green development.

Keywords: artificial intelligence; green total factor productivity; green innovation; industrial structure upgrading; land resource misallocation

1. Introduction

At present, China's economic development is facing increasingly strong resource and environmental constraints. Moreover, the production function, combination methods of various factors, and allocation efficiency in economic development are undergoing changes. Dual carbon targets have become an important framework for China's economic development, and the green transformation of development methods has become a necessary path under multiple constraints [1]. The key to the green transformation of China's economy lies in gradually relying on technological innovation and production transformation to transform into a green production model with low emissions, low energy consumption, and high efficiency, while timely abandoning the traditional extensive production model to achieve green economic development [2]. Although most scholars use TFP to measure economic development, this approach does not consider the impact of factors such as energy and environment. Moreover, the measurement of GTFP includes the aforementioned variables, which is consistent with current economic development concepts. GTFP is an important driving factor for accelerating the transformation of driving forces and promoting sustainable economic development. GTFP also has important research value. Hence, improving GTFP is a necessary path to achieving the dual carbon goals [3].

At present, the academic community is delving into the driving factors of GTFP from different dimensions, such as industrial structure, external investment, environmental regulations, digital economy, and industrial policies. However, the core driving force behind the improvement of GTFP is technological innovation, especially strategic technological advancements, such as AI. In this context, various countries or regions have placed the development of AI in an important strategic position, formulated AI development strategies, and maximized the initiative and discourse power of AI development to seize the opportunity in the new round of international industrial division of labor adjustments and interest games. The Chinese government has attached immense importance to the promoting role of AI in the technological revolution, particularly emphasizing its role in stimulating high-quality economic development. To further promote the development of AI, China implemented a pilot policy called AIPZ in 2019, bringing new opportunities for improving GTFP. The AIPZ focuses on promoting the deep integration of AI and economic development. This undertaking aims to form a batch of replicable and promotable experiences and comprehensively enhance the innovation ability and level of AI. AI can provide accurate environmental data and resource information, providing a basis for the government to formulate green development plans. In addition, AI supports the intelligent and refined management of cities, thereby improving resource utilization efficiency.

Given the preceding discussion, the following critical questions must be answered. What is the relationship between AI and GTFP in the current trend of intelligent transformation? What is the mechanism of its impact? The existing literature does not provide exact answers to these questions. On the basis of the AIPZ policy implementation, we use the progressive difference-in-differences (DID) model and combine relevant urban data from 2010 to 2021 to analyze the impact and heterogeneity of AI on GTFP. The objective is to evaluate the impact of government-driven AI development policies on GTFP. In the mechanism analysis section, the current research utilizes existing research conclusions as a basis to analyze the mechanism of AI on GTFP from industrial structure upgrading (ISU), green innovation (GI), and land resource misallocation (LRM). Moreover, this study intends to conduct empirical testing. On the one hand, this study further affirms the importance of ISU, GI, and LRM in promoting green development. On the other hand, this research provides a theoretical framework for subsequent related research. Figure 1 shows the research approach.

The research contribution is as follows: First, we test the effect of AI on GTFP based on the AIPZ pilot policy. This study also emphasizes the role of AI in green development, providing beneficial ideas for exploring the path of green ecological construction in China. Second, the existing literature usually uses single indicators, such as industrial robots, industrial robot application density, and a number of AI patents to measure. The advantages of these indicators lie in their clear connotation and ease of data collection. However, their disadvantage is that they are one-sided and may underestimate or overestimate the comprehensive development level of AI, thereby interfering with empirical results. This study utilizes the exogenous impact of the regional AIPZ pilot policy to construct a quasi-natural experiment. In particular, the progressive DID method is used to alleviate the bias caused by indicator selection and accurately explore the causal effect between AI and GTFP. The goal is to obtain markedly convincing conclusions. Third, this research studies the impact mechanism of AI on urban GTFP, including the effects of ISU, GI, and LRM. This aspect reveals the "mechanism black box" of AI that empowers China's green transformation. This finding is conducive to enriching sustainable development theory and provides basic literature for subsequent research. The research conclusions can provide empirical support for AI development, further expand the pilot scope of AIPZ, and solidify the implementation path for addressing climate change challenges.

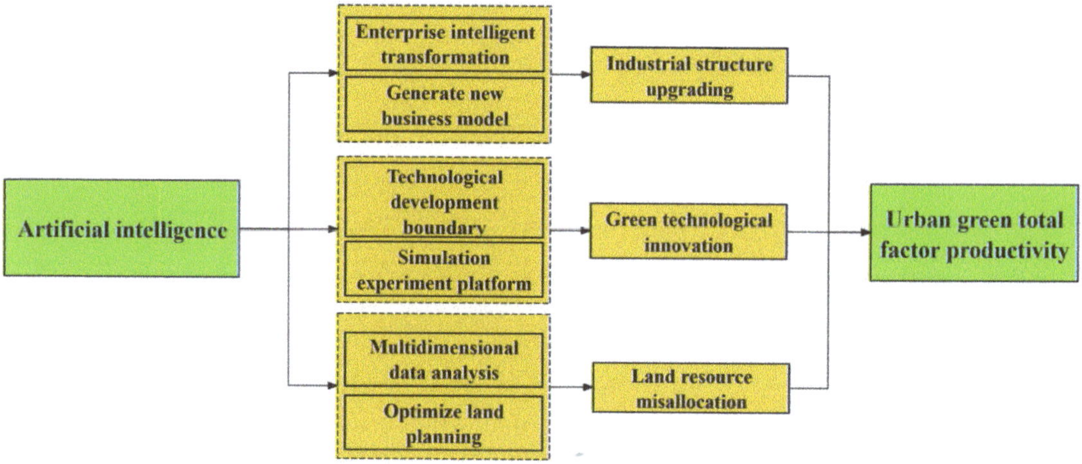

Figure 1. Research approach.

2. Literature Review

2.1. Research on AI

AI has extensive connotations. From a technical perspective, AI is the cognitive science of intelligent computer programs and belongs to a branch of computer science [4,5]. From the perspective of application connotation, AI, as a disruptive technological innovation, has significant interdisciplinary and extensive penetration. The academic community views AI as a general-purpose technology similar to steam engines and computers. Moreover, AI is a common technological advancement and also a technological advancement with infrastructure properties, fully penetrating into various industries and profoundly changing industrial development models and factor endowment structures [6–9]. The academic community mainly measures AI from four dimensions. The first three dimensions are the application of industrial robots [10–12], AI technology patents [13], and the development level of the AI industry [14,15]. For the fourth dimension, scholars construct a multidimensional indicator system to measure a comprehensive index of AI from such dimensions as software popularization and application, intelligent manufacturing industry, and industrial enterprise innovation capabilities [16].

The economic effects of AI are mainly reflected in economic growth, labor employment, and income distribution. Academic research has found that AI mainly promotes economic growth through such mechanisms as improving labor quality, capital return rate, and TFP [17–20]. AI may have two completely different impacts on the labor market. The first impact is the substitution effect, which means that AI leads to "machine replacement". The second impact is the creation effect, which means that the development of AI will generate new industries, formats, and products, among others, which will enhance the market's demand for highly skilled labor and create new job opportunities. The most significant challenge faced by the development of AI is income distribution, which is mainly manifested in the income gap caused by the income distributions of different factors and heterogeneous workers [21,22]. At present, the research on the environmental effects of AI is limited. A few scholars have studied the effects of AI on GI and pollution reduction and found that AI can bring positive environmental effects [23,24]. The research on how AI affects regional GTFP has not yet received attention.

2.2. Research on GTFP

TFP is defined as output attributed to technological progress and efficiency improvement in economic growth, in addition to the increase in factor input, which can be used to reflect the quality of economic growth [25]. With the continuous intensification of global environmental pollution, the importance of environmental regulation is becoming increasingly prominent, and sustainable development is receiving increasing attention. Research on green development is also increasing. GTFP incorporates unexpected output into the indicator system and considers environmental factors [26]. The most commonly used method for calculating GTFP is data envelopment analysis (DEA). DEA considers non-expected outputs, such as pollution emissions, which have a negative impact on resources and the environment, as output variables and combines them with the directional distance function to process the non-expected and measure the GTFP value. Research on the influencing factors of GTFP mainly focuses on economic growth, digitization, green finance, and environmental regulation. Erkul and Türköz (2024) found that economic growth is negatively related to GTFP [27]. Wu et al. (2020) determined that environmental regulations will first suppress urban GTFP and then improve GTFP [28]. Guo et al. (2023) indicated that green finance can improve GTFP [29]. With the popularization of information networks and the continuous advancement of AI in China, scholars have also studied the effect of data-driven on GTFP. Gao et al. (2022) found that information technology can significantly improve GTFP [30]. Lyu et al. (2023) determined a correlation between the digital economy and GTFP [31].

The existing literature has also begun to focus on the relationship between AI and GFTP, and empirical tests have been conducted based on macro- and micro-level data in China. The results show that AI has a positive impact on GTFP [32,33]. However, these studies did not explore the impact of AI on GTFP from a quasi-natural experimental perspective. Owing to the late implementation of China's AIPZ policy, there is currently limited empirical testing of the effects of AIPZ's policy implementation. Therefore, examining the effectiveness of AIPZ policy on urban GTFP and studying the environmental effects of AI have stronger scientific and practical significance. This finding provides empirical support for further improving and promoting the use of AI policies.

3. Policy Background and Research Hypothesis

3.1. Policy Background

As early as 2015, China has proposed to include AI in the national priority development area and has included the development of AI in government work reports for several consecutive years. To comprehensively enhance the innovation capacity and level of AI, the Chinese government issued the "Guidelines for the Construction of the National New Generation Artificial Intelligence Innovation and Development Pilot Zone" in 2019 for the orderly promotion of the AIPZ construction [34]. Beijing took the lead in establishing an AI pilot zone, followed by Shanghai. As of 2022, China has approved 18 cities to establish AI pilot zones. The implementation time of the pilot cities is shown in Table 1 below. The selection criteria for pilot cities are mainly based on their abundant scientific and educational resources, good AI industry tutorials, relatively complete network infrastructure, and strong policy support from local governments for the development of AI [34]. The AIPZ policy has four main requirements. The first requirement is to increase the R&D effort of basic theories and conduct research and application demonstrations of AI technology in manufacturing, home furnishings, medicine, and other fields. The second requirement is to carry out policy pilot efforts around data openness and protection, achievement transformation, intellectual property, security management, and other aspects; and to form a policy framework and regulatory standards system that adapts to AI development. The third requirement is to strengthen the accumulation of social experimental theories, accurately identify AI challenges, comprehend the laws of social evolution in the AI era, and improve the precision and scientific level of government governance in the intelligence era. The fourth requirement is to strengthen AI construction, such as communication networks, big data centers, and computing centers, and to form a high-level infrastructure system supporting the AI applications.

Table 1. Implementation timeline of pilot cities.

	Pilot City	Policy Implementation Time
1	Beijing	2019
2	Shanghai	2019
3	Tianjin	2019
4	Shenzhen	2019
5	Hangzhou	2019
6	Hefei	2019
7	Huzhou	2019
8	Chongqing	2020
9	Chendu	2020
10	Xian	2020
11	Jinan	2020
12	Guangzhou	2020
13	Wuhan	2020
14	Suzhou	2021
15	Changsha	2021
16	Zhenzhou	2021
17	Shenyang	2021
18	Haerbin	2021

According to the Chinese government requirements, each pilot zone relies on regional advantages, innovative resources, and distinctive industrial foundations to vigorously build an AI infrastructure. Moreover, the pilot zones play an important demonstration role in promoting local high-quality development. So, the following question is crucial: Can the AIPZ pilot policy promote the improvement of GTFP?

3.2. *Research Hypothesis*

3.2.1. AI Affects GTFP Through ISU

The innovative application of AI technology promotes the vertical development of industrial intelligence, changes traditional economic activities, and becomes an important engine for the ISU [35]. Firstly, enterprises are the main demanders of factors and the main implementers of economic benefits. By introducing or innovating their own R&D, enterprises obtain intelligent production equipment and put it into production to achieve their own intelligent transformation and improve their production efficiency [36]. AI technology can promote the transformation of traditional elements into data, using training data for deep learning to achieve value creation, and making production and manufacturing data play an increasingly important role in industrial construction. In addition, the transformation of intelligent production will promote the formation of new business models, encourage the transfer of production factors in the national economy to higher productivity and value-added industrial sectors, and fundamentally optimize the industrial structure system [37], achieving a high-level transformation of the industrial structure.

As an important link between resources, environment, and economic development, the ISU is crucial for improving urban GTFP [38,39]. ISU can promote efficient allocation of resources, shifting from inefficient polluting sectors to efficient clean sectors, reducing resource waste and environmental burden, and enhancing green productivity [40]. The spillover effects of technological progress are significant in industrial transformation. Technology-intensive industries accelerate the transformation of traditional industries to emerging green industries through technology diffusion and industrial linkage and overall enhance GTFP. Therefore, AI can improve GTFP by improving ISU.

3.2.2. AI Affects GTFP Through GI

The development of AI has driven breakthroughs in green technology for enterprises, extending their technological development boundaries in environmental pollution prevention and control [41]. On the one hand, AI applications help enterprises achieve efficient green technological innovation. GI often requires trial and error, and the development of AI provides a simulated experimental platform for enterprises' GI. Enterprises can invest in intelligent devices and apply cloud computing platforms to conduct multiple rounds of GI research and development with different solutions, correct and improve future GI paths based on existing historical trial and error data, and expand the efficiency of internal resource allocation and spatial storage capacity of enterprises to ensure that they can achieve predictable GI in an efficient manner.

AI has promoted the application of clean technology in enterprises. Clean energy sources have evident environmental protection effects, but the use of these types of energy requires the assistance of technical equipment and talent resources [42–44]. AI applications can provide enterprises with existing mature clean energy investment plans and based on the collection and iteration of backend big data, provide accurate technical monitoring and energy investment ratio calculation at all times. This situation improves the application of clean technology, ultimately achieving the development of GI. The enhancement of GI capabilities can further improve enterprises' energy consumption structure in the production process, thereby reducing the scale of investment in traditional polluting energy [45]. GI can improve energy efficiency, thereby reducing the formation of pollution by-products from the front-end control of enterprise production, reducing pollution emission intensity, and enhancing GTFP [46,47]. Therefore, AI can improve GTFP by improving GI.

3.2.3. AI Affects GTFP Through LRM

AI technology provides efficient and accurate solutions for the investigation and evaluation of land resources through technologies such as big data, remote sensing, and geographic information systems [48]. In the planning and optimization of land resources, AI technology can provide multi-dimensional data analysis and intelligent decision support, helping decision-makers to better formulate land use planning and optimize allocation schemes. Through AI technology, government departments can simulate different utilization plans for land resources and optimize them through algorithms to find the optimal land use plan, achieving optimal allocation of land resources, increasing expected output, and reducing LRM.

The LRM will lead to the continuous expansion of urban space toward the suburbs, and the supply of cheap industrial land will result in the emergence of a large number of development zones, industrial areas, and new cities around the city. The unsuitability of such land resources has led to the extensive use of urban land, exacerbating the disorderly expansion of cities [49]. The disorderly expansion of cities is not conducive to economic agglomeration and economies of scale, making urban economic activities more dispersed. Urban sprawl can lead to a decrease in population and employment density, negative impacts on the use of infrastructure and knowledge spillovers, weaken urban competitiveness, and reduce urban green economic efficiency [50]. LRM can also damage the green spaces around the city, reduce the local ecological regulation function, and lead to an increase in carbon dioxide and pollutant emissions, which is not conducive to the green development of the city [51]. In addition, LRM can squeeze the land resources of emerging enterprises, causing excessive tilt of land resources toward inefficient industrial enterprises, squeezing the land space of high-tech enterprises, and inhibiting resource investment in green technology innovation [49]. At the same time, low-cost land weakens the motivation of enterprises to reduce pollution through technological innovation, resulting in insufficient motivation for technological upgrading and hindering the overall improvement of GTFP [50,51]. Therefore, AI can reduce the degree of LRM, thereby improving the overall GTFP of cities.

In summary, we propose the following research hypothesis: AI mainly affects GTFP by improving ISU and GI and reducing LRM. Figure 1 shows the research approach.

4. Research Design

4.1. Model

This study regards the AIPZ policy implemented in China in 2019 as an exogenous shock to test the effect of AI on GTFP. Due to the fact that the approval of pilot cities is not one-time, but phased, there are significant differences in the specific years when the experimental group cities are selected as demonstration cities. This progressive approval method also makes it difficult for us to set a unified virtual variable for the sample opening time point like the traditional DID method. We employ a progressive DID method to examine whether AI can improve GTFP. This paper uses urban-level data for analysis, and the data come from CNRDS. The progressive DID model is set as follows:

$$GTFP_{it} = \alpha_0 + \alpha_1 DID_{it} + \alpha_2 Conl_{it} + City_i + Year_t + \varepsilon_{it}, \quad (1)$$

GTFP denotes the GTFP of cities. *DID* denotes the AIPZ *pilot* policy. If city i is selected in year t, then DID_{it} is 1, otherwise it is 0. *Conl* is the set of control variables (CV), $City_i$ is the city-fixed effect, and $Year_t$ is the year-fixed effect.

4.2. Variable

GTFP: Referring to the existing literature [20,52], the super-efficiency slack-based measure-undesirable (SBM) model is used to calculate the GTFP index. In terms of input and output indicators, the labor, capital, and total electricity consumption are selected as inputs [52]. The gross regional product, industrial sulfur dioxide, smoke, and wastewater are selected as undesirable outputs.

CV: Considering that other urban characteristic factors may have an impact on urban GTFP, drawing on relevant research [28–30], this study controls for the following variables: economic development (EC), city size (CS), urbanization level (UL), financial development (FIN), government intervention (GIN), infrastructure construction (IC), opening up (OU). Their calculation methods are shown in Table 2.

Table 2. Descriptive statistics.

Variable	Obs	Definition	Mean	SD
GTFP	2974	GTFP index based on super-efficiency SBM model	0.602	0.230
EC	2974	The logarithm of per capita GDP	10.705	0.552
CS	2974	The logarithm of the total population	8.112	1.587
UL	2974	Total urban population/total population	0.521	0.146
FIN	2974	The logarithm of the balance of deposits in financial institutions	14.423	2.807

Table 2. *Cont.*

Variable	Obs	Definition	Mean	SD
GIN	2974	The general public budget expenditure/GDP	0.198	1.159
IC	2974	The logarithm of per capita road area	1.737	4.431
OPEN	2974	The logarithm of the proportion of foreign direct investment to GDP	−6.411	1.226

5. Results

5.1. Benchmark Results

Table 3 reports the impact of AI on GTFPP. The *DID* coefficients from columns (1) to (3) indicate that AI has improved the GTFP of cities, which is consistent with the research hypothesis. The reliability of the results in this study still needs further verification, so we will conduct a series of robustness tests to ensure the reliability of the main conclusions.

Table 3. The effect of AI on GTFP.

	(1) GTFP	(2) GTFP	(3) GTFP
DID	0.0834 ***	0.228 ***	0.0845 ***
	(0.0180)	(0.0448)	(0.0202)
EC		0.245 ***	−0.0130
		(0.0211)	(0.0261)
CS		−0.0114	−0.0480 *
		(0.00720)	(0.0252)
UL		−0.0908 ***	0.0876
		(0.0328)	(0.0545)
FIN		0.0663 **	−0.0419
		(0.0280)	(0.0270)
GIN		−0.182 ***	−0.238 ***
		(0.0187)	(0.0250)
IC		−0.0260 ***	−0.00715
		(0.00462)	(0.00850)
OPEN		−0.00114	−0.0118
		(0.00790)	(0.0190)
City Effect	NO	NO	YES
Year Effect	NO	YES	YES
Observations	2974	2974	2974
R-squared	0.771	0.193	0.787

Note: *, **, *** represent significance levels at 10%, 5%, 1%, respectively, with robust stand errors in parentheses.

5.2. Robustness Test

5.2.1. Parallel Trend

Drawing on the event study method, annual dummy variables are set up separately by sample year to interact with group dummy variables and thus observe the time trend of the average treatment effect. Table 4 displays the parallel trend results. Before the implementation of the AIPZ policy, the interaction coefficients (Before1 and Before2) are not significant, which means that there is no significant difference in the trend of GTFP changes between pilot and non-pilot cities. After the implementation of the AIPZ policy, the coefficients of the interaction term (Current, Aftrer1, and After2) are significant, indicating a significant difference in the trend of GTFP changes between pilot and non-pilot cities, and providing supporting evidence for the parallel trend hypothesis. To visually demonstrate the changing trend of policy effectiveness, we use Figure 2 for illustration.

Table 4. Parallel trend test.

	(1) GTFP	(2) GTFP
Before2	0.00929	0.00254
	(0.0214)	(0.0223)
Before1	0.0286	0.0214
	(0.0259)	(0.0262)
Current	0.117 ***	0.110 ***
	(0.0274)	(0.0253)
After1	0.129 ***	0.121 ***
	(0.0256)	(0.0277)
After2	0.0804 ***	0.0788 ***
	(0.0269)	(0.0288)
Control	NO	YES
City Effect	YES	YES
Year Effect	YES	YES
Observations	2974	2974
R-squared	0.772	0.788

Note: *** represent significance levels at 1%, with robust stand errors in parentheses.

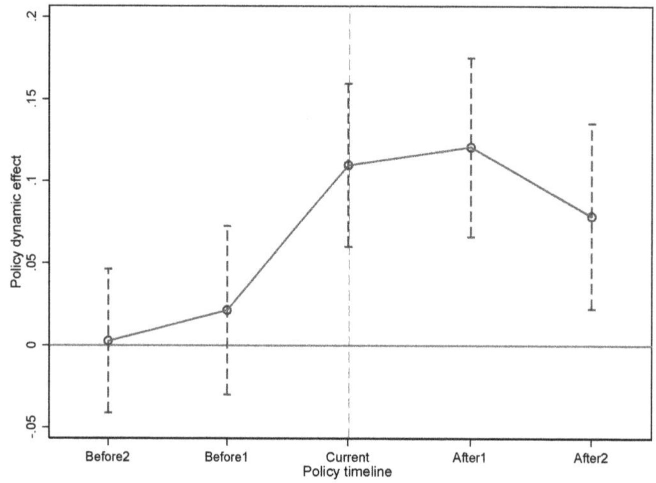

Figure 2. Parallel trend.

5.2.2. Sensitivity Analysis of Parallel Trend

The parallel trend assumption is the key to the ability of DID estimation to cleanly identify the average treatment effect. It assumes that the difference in mean results between the treatment and control group will continue after treatment without any processing occurring. However, the parallel trend hypothesis is untestable, and researchers typically indirectly validate its validity through the pre-treatment trend test and co-shock test. The sensitivity test method proposed by Rambachan and Roth (2023) [53] evaluates the violation of the parallel trend by setting constraints, thereby providing more robust inferences. This study refers to Biasi and Sarsons (2022) [54] to conduct a parallel trend sensitivity analysis on the impact of GTFP in the first and second years after the event occurred. Figures 3 and 4 show the parallel trend sensitivity test results of the treatment effects of AI on GTFP using the Bounds on Relative Magnitude method. Figures 5 and 6 show the parallel trend sensitivity test results of the treatment effects of AI on GTFP using the Smoothness Restriction method. As shown in Figures 3 and 4, whether in the first or second year of implementation of AIPZ policy, within a reasonable range of deviation from the parallel trend, the core conclusion of this paper still holds true.

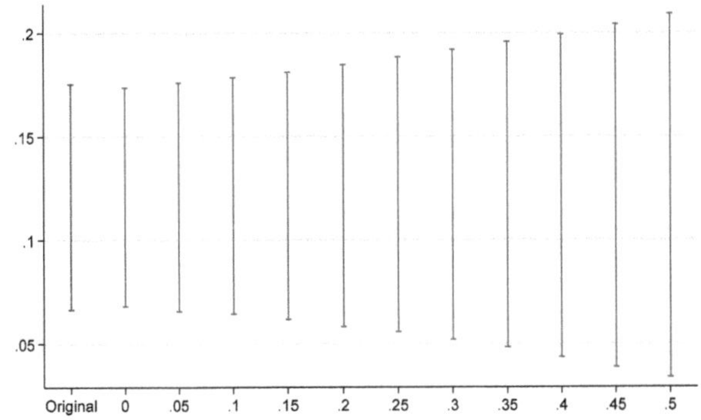

Figure 3. Bounds on relative magnitude test for After1.

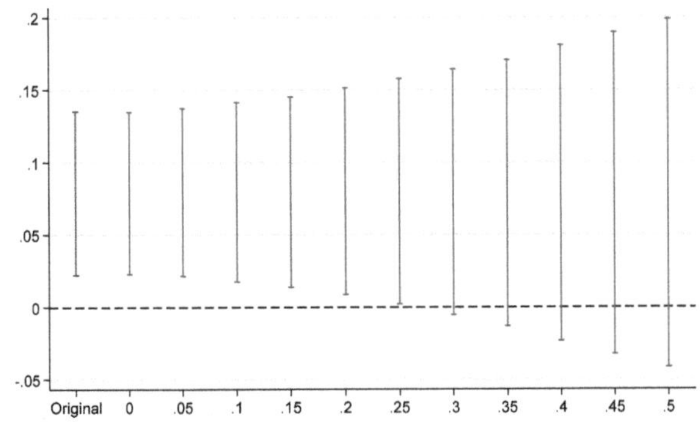

Figure 4. Bounds on relative magnitude test for After2.

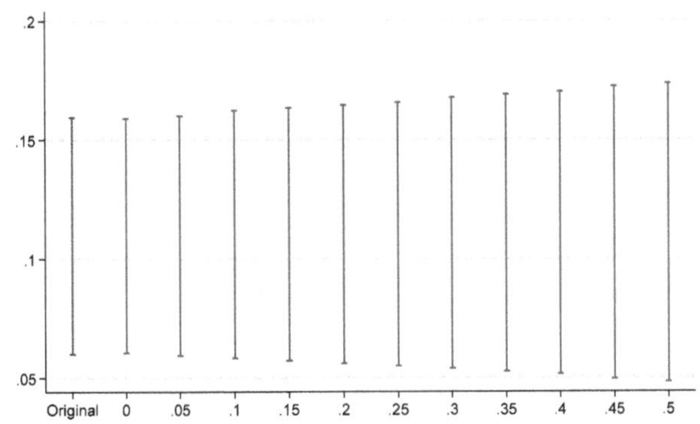

Figure 5. Smoothness restriction test for After1.

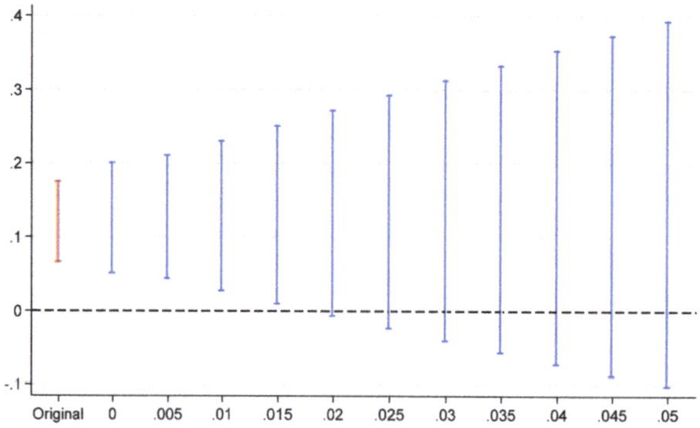

Figure 6. Smoothness restriction test for After2.

5.2.3. Heterogeneity Treatment Effect

Currently, scholars have discussed the heterogeneous treatment effects generated by the DID method. The two-way fixed effects model (TWFE) can obtain unbiased estimates of the true treatment effects when dealing with homogeneity of effects. However, if there is heterogeneity in the treatment effects between groups or when they are heterogeneous, the TWFE regression results may yield estimates of treatment effects that are difficult to explain [55]. At this point, the estimated coefficients may not be the convex weighted average sum of various treatment effects, ultimately affecting the estimation of regression coefficients. To solve this problem, we refer to the treatment method of de Chaisemartin and D'Hautfoeuille (2020) [55] and calculate the instantaneous processing effects of AI on GTFP in different relative periods. In Figure 7, we find that the DID coefficient remained small before the event but shows a significant increase and significance in the first period after the event. Overall, after considering the heterogeneity of treatment effects, the basic results of this study do not show significant bias.

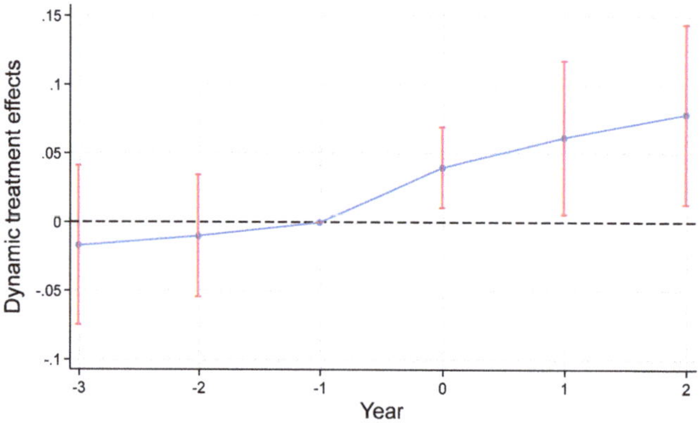

Figure 7. Heterogeneity treatment effect.

5.2.4. Placebo Test

Due to some unobservable variables that may simultaneously affect AI and GTFP, we use a placebo test to mitigate this adverse effect. The placebo test examines the reliability of policy effects by using a fake experimental group. We follow the approach of La Ferrara et al. (2012) [56] and randomly sample 500 times. We then re-estimate the DID coefficients and P-value distribution using model (1), as shown in Figure 8. The mean regression coefficients of GTFP are close to 0, and the p-values are mostly greater than 0.1. Therefore, the impact of AI on GTFP is not caused by other random factors, and the conclusion drawn in the previous text is reliable.

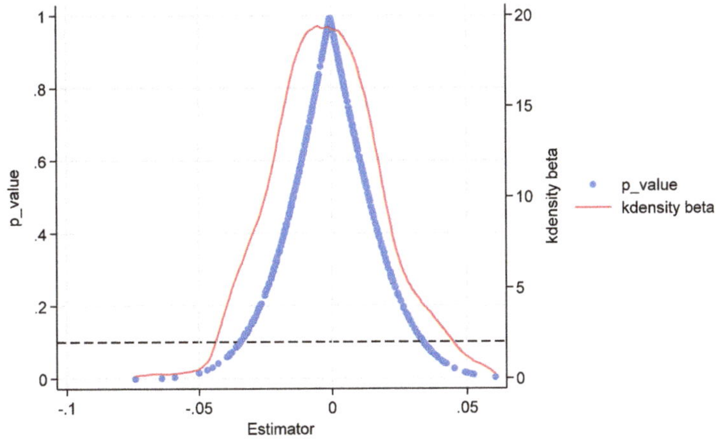

Figure 8. Placebo test.

5.2.5. PSM-DID Estimation

Although the DID method separates the average treatment effect of the pilot policy, there may still be selective bias in observing and researching data due to the fact that the AIPZ policy is not a strictly natural experiment. We use the propensity score matching (PSM) method to alleviate this problem. Referring to Zhou and Qi (2024) [57], this study uses the control variables as the identification feature of the sample points to perform the nearest neighbor, kernel, and radius matching between demonstration and non-demonstration cities. The DID method is then used to regress the matched sample. Table 5 displays the results of PSM-DID, and the previous findings remain consistent.

Table 5. PSM–DID.

	(1) **Nearest Neighbor** **GTFP**	(2) **Kernel** **GTFP**	(3) **Radius** **GTFP**
DID	0.043 ***	0.0923 ***	0.119 ***
	(0.0131)	(0.021)	(0.024)
Control	NO	YES	YES
City Effect	YES	YES	YES
Year Effect	YES	YES	YES
Observations	1336	2971	2932
R-squared	0.417	0.788	0.781

Note: *** represent significance levels at 1%, with robust stand errors in parentheses.

5.2.6. Other Robustness Tests

The existing literature uses the level of innovation in AI technology to measure the development level of AI. We use AI patents to measure AI. We collected AI patent application data from various prefecture-level cities [13], logarithmically processed it, and used it as the independent variable for regression analysis in this study. Column 1 in Table 6 displays the results of replacing the independent variable. The DID estimation result is significant. Considering that the control variables may have a reverse causal relationship with the dependent variable and affect the results of this study, we have lagged the control variables by one period. According to the second column of Table 6, it is found that the main result of this study does not change when the control variables are lagged by one period and added to Model (1). The ideal scenario for evaluating the effect of AI on GTFP using the DID method is that the pilot cities are randomly selected. However, in reality, the selection of pilot cities is often based on factors such as the city's economic development, urbanization level, financial development, and infrastructure construction. The inherent differences in these cities will have varying impacts on the GTFP over time, leading to estimation errors. To control the impact of the non-random selection problem of AI pilot cities, we draw on the approach of Li et al. (2016) [58] and add an interaction term between control variables and time trends on the basis of benchmark regression. The results of column 3 show that after considering the non-random selection problem of pilot cities, the core coefficient is still significant.

Table 6. Other robustness tests.

	(1) **GTFP**	(2) **GTFP**	(3) **GTFP**
DID	0.135 ***	0.074 ***	0.087 ***
	(0.044)	(0.019)	(0.02)
Control	YES	YES	YES
City Effect	YES	YES	YES
Year Effect	YES	YES	YES
Observations	2974	2521	2974
R-squared	0.855	0.809	0.791

Note: *** represent significance levels at 1%, with robust stand errors in parentheses.

5.2.7. Robustness Test Based on GRF

The GRF model can randomly divide subsamples by recursion, calculate the processing effect at the overall sample observation, ensure the effectiveness of the overall regression results, and determine the weight of the processing effect based on data characteristics. Thus, this approach avoids the problem of artificial settings and interference. Specifically, based on the research of Wager and Athey (2018) [59], we use the "honest" method to construct the following GRF model for estimating the average processing effect of AI on GTFP. The average treatment effect of the policy estimation of the GRF model is shown in Table 7. Columns 1–4 present the results of the GRF model. When the number of decision trees increased from 500 to 3000, the average treatment effect of AI on GTFP remained stable at around 0.205, which supports the impact of AI in improving GTFP.

Table 7. GRF results.

	(1) **GTFP**	(2) **GTFP**	(3) **GTFP**	(4) **GTFP**
DID	0.204 ***	0. 205 ***	0. 205 ***	0. 205 ***
	(0.031)	(0.029)	(0.028)	(0.028)
Tree Num	500	1000	2000	3000
Model	Causal Forest	Causal Forest	Causal Forest	Causal Forest
Observations	2974	2974	2974	2974

Note: *** represent significance levels at 1%, with robust stand errors in parentheses.

5.3. Mechanism Analysis

We construct the Model (2) to examine the mechanism by which AI affect GTFP:

$$Med_{it} = \beta_0 + \beta_1 DID_{it} + \beta_2 Conl_{it} + City_i + Year_t + \varepsilon_{it} \quad (2)$$

Med is the mechanism variable, representing GI, ISU, and LRM, respectively. We use the logarithm of the number of urban green patent applications as a proxy variable for GI [60]. We use the ratio of the tertiary industry value to the secondary industry as a proxy variable for ISU [57]. The overall allocation of urban construction land in China is characterized by a large supply of industrial land while relatively restricting the supply of commercial and residential land, leading to the expansion of industrial land and a relative shortage of commercial and residential land. This land allocation model not only affects the efficiency of urban land use but also leads to the misallocation of urban construction land. Based on An (2024) [61], this paper measures the LRM by the proportion of urban industrial land area to construction land area. The higher the proportion, the larger the industrial land area. Under a certain total urban land volume, the expansion of industrial land will inevitably occupy the commercial and residential land area. Therefore, this indicator can better characterize the distortion and mismatch of the supply structure of industrial land and commercial and residential land.

Table 8 reports the intermediate effect test. The DID coefficients in column (1) and column (2) are positive, indicating that AI can improve GI and ISU. The DID coefficient in column (3) is negative, which means that AI can reduce LMR. The existing literature has found that GI and ISU have a significant effect on enhancing GTFP, while LRM can reduce GTFP [38,42–47,51]. Therefore, GI, ISU, and LMR are the intermediate variables for AI to improve GTFP. To increase the robustness of the results, we once again use the GRF model to validate the impact of AI on GI, ISU, and LRM. As shown in Table 9, the results estimated using the GRF model still show that the AI mainly affects GTFP through GTI, ISU, and LRM.

Table 8. Mechanism test.

	(1) **GI**	(2) **ISU**	(3) **LRM**
DID	0.054 *	0.142 ***	−0.022 *
	(0.0315)	(0.041)	(0.012)
Control	YES	YES	YES
City Effect	YES	YES	YES
Year Effect	YES	YES	YES
Observations	2974	2974	2974
R-squared	0.801	0.891	0.832

Note: *, *** represent significance levels at 10%, 1%, respectively, with robust stand errors in parentheses.

Table 9. GRF results.

	(1) **GI**	(2) **ISU**	(3) **LRM**
DID	0.121 ***	0. 338 **	−0. 017 ***
	(0.04)	(0.154)	(0.004)
Tree Num	3000	3000	3000
Model	Causal Forest	Causal Forest	Causal Forest
Observations	2974	2974	2974

Note: **, *** represent significance levels at 5%, 1%, respectively, with robust stand errors in parentheses.

5.4. Heterogeneity

5.4.1. Heterogeneity of Urban Human Capital

Some uncertainty remains about whether the research and application of GI can be gradually promoted in pilot cities, and good human capital (HC) can help reduce this uncertainty. The spatial agglomeration of HC helps to accumulate and innovate GI, creating conditions for existing technological innovation and research and development. The spillover characteristics of HC can also effectively promote the promotion and application of GI, thereby further reducing environmental pollution. This study uses the ratio of the number of students in ordinary higher education institutions to the total population of the city to represent the level of urban HC. We assign a value of 1 to cities with high HC levels, otherwise, it is 0. We add the interaction term (HC_did) between the HC and AI to Model 1 for regression. The regression coefficient of *HC_did* in Table 10 is significantly positive, indicating that the AI can significantly improve the degree of GTFP of high HC-level cities compared with low HC-level cities. The reason may be that the higher the level of urban HC, the more obvious the advantage of talent concentration, and the more effective AI policy can play in optimizing the allocation of urban resource elements, thus accelerating the integration and interaction between AI and local industries, promoting industrial optimization and upgrading, and effectively promoting the increase in GTFP.

Table 10. Heterogeneity test.

	(1) GTFP	(2) GTFP	(3) GTFP
HC_did	0.037 * (0.02)		
Resource_did		0.021 * (0.011)	
GCB_did			0.033 ** (0.016)
Control	YES	YES	YES
City Effect	YES	YES	YES
Year Effect	YES	YES	YES
Observations	2974	2974	2974
R-squared	0.788	0.787	0.788

Note: *, ** represent significance levels at 10%, 5%, respectively, with robust stand errors in parentheses.

5.4.2. Heterogeneity of Urban Resource Type

The transformation of traditional energy-based cities is an objective requirement for the sustainable development of cities, and excessive reliance on traditional energy development models can hinder their urban transformation. The differences in urban resource types may affect the execution of energy transformation policies by local governments. The heterogeneous impact of urban resource type differences on the policy effect of AI policy must be studied. Referring to Zhou and Qi (2022) [52], we divide the sample cities into resource-based (RB) cities and non-resource-based (NRB) cities. We define NRB cities as 1 and RB cities as 0, and add the interaction term (Resource_did) between the urban resource type and AI to Model 1 for regression. The coefficient of *Resource_did* in Table 10 is positive, indicating that the AI can improve the GTFP of NRB cities compared with RB cities.

RB cities are influenced by local natural resource endowments, mainly focusing on traditional heavy industry as the direction of urban economic development. Long-term excessive resource consumption and environmental pollution are difficult to repair and adjust, and the establishment of an AI pilot has little effect on improving urban GTFP. NRB cities are less constrained by the traditional heavy industry economy and also pay attention to the cultivation and absorption of digital technology and talents. The approved construction of the AI pilot can to some extent guide these cities to accelerate the development of green and digital industries, which has a significant driving effect on GTFP.

5.4.3. Heterogeneity of Urban Green Consumption Behavior

Green consumption behavior (GCB) refers to consumers choosing green products that are not polluted or contribute to public health in social consumption. GCB focuses on conserving resources and energy, achieving sustainable consumption, and avoiding or reducing environmental damage during the consumption process. It is a new type of consumption behavior and process that advocates for nature and ecological protection. We select the logarithm of the total number of passengers carried by public buses and trams per year to measure GCB [57]. We assign a value of 1 to cities with high GCB levels, otherwise, it is 0. We add the interaction term (GCB_did) between the GCB and AI to Model 1 for regression. The *GCB_did* coefficient in Table 10 is positive, indicating that the AI can improve the GTFP of high GCB-level cities compared with low GCB-level cities.

The development of AI has provided new possibilities for achieving green consumption. AI can promote the upgrading of traditional consumption demand, create green and intelligent consumption, comprehensively improve consumption quality, and continuously meet the growing needs of people for a better life [44]. Moreover, upgrading green consumption demand has forced industries' green development. AI can analyze large-scale consumption data and predict the demand for green products to help manufacturers and retailers better meet consumer expectations. AI can also provide consumers with personalized green product recommendations based on personal consumption history and preferences, encouraging them to make environmentally friendly choices. In addition, AI can help consumers better understand the importance of green consumption by providing educational content and behavioral feedback to guide them to change traditional consumption behavior [44]. Therefore, in areas with a high level of GCB, AI can better promote industrial upgrading, thereby enhancing green GTFP.

6. Conclusions and Policy Implications

As the world's largest developing country and carbon emitter, China is facing urgent environmental pressures and urbanization challenges. GFTP can more comprehensively and realistically reflect the economic development status of cities considering environmental impact and resource utilization efficiency. In this context, studying the impact of the development of AI on urban GTFP has important theoretical and practical significance for exploring new economic growth drivers and promoting sustainable economic and social development. We employ a progressive DID model to test the effect of AI on GTFP. The research findings indicate that AI significantly improves GTFP, the conclusion that holds even after robustness tests. Mechanism analysis suggests that AI enhances GTFP by promoting ISU, improving GI, and reducing LRM. Heterogeneity analysis reveals the policy effect is more significant in high HC-level cities, NRB cities, and high GBC cities. Our main research findings are consistent with the conclusions of the existing literature [32,33], discovering the positive impact of AI on GFTP and enriching the research boundary on the environmental effects of AI. The research conclusions provide data support and a new perspective for evaluating the environmental effects of AI development and provide a reference for exploring the improvement path of GTFP and the formation of corresponding new quality productive forces. In terms of policy implications, according to the main conclusions of theoretical analysis and empirical evidence, this discussion puts forward relevant suggestions from the perspectives of formulating relevant policies to improve the positive effects of AI and encouraging enterprises to introduce AI.

First, the conclusion of this paper indicates that the development of AI generally promotes the GTFP of cities. Therefore, policymakers should further promote the widespread application of AI in various industries to maximize its positive effects. National and local governments should strengthen guidance and supervision on the construction of AIPZ cities, improve the evaluation system of pilot cities, fully leverage the positive role of pilot cities in optimizing energy structure and promoting ecological civilization, and promote the accelerated development of AI.

Second, on the one hand, application and development of AI technology should be encouraged, so that the AI industry can form a virtuous circle from research and development to application. On the other hand, encouraging the active enhancement of the hard and soft power that matches the development of AI technology seems important, including further promoting the construction of new infrastructure, improving the integrity of the supply chain, and increasing talent and technical reserves to ensure that relevant factors and supply chain match the development of AI. Meanwhile, measures like actively advancing job guidance and job training to help the workforce acquire skills complementary to AI technology, as well as reducing the gap between the rich and the poor should be taken so that more people can benefit from the popularization of AI.

Third, firms should recognize their own positioning, reasonably introduce AI into the advanced technology of the production processes instead of blindly promoting advanced technology, and take advantage of the first-mover advantage of AI and its more obvious effects on low-quality products to gain more benefits from intelligent production. In addition, enterprises should increase green technical innovation to convert the cost advantages of AI into technical advantages, so as to avoid wasting the opportunity presented by AI.

Our research still has certain limitations. There may be some controversy in the measurement of some variables, such as in the measurement of GCB and HC. Due to limitations in data availability and references, the proxy indicators we choose may not fully reflect their true situation. We will further collect relevant data in the future and strive to have more comprehensive measurements for each variable. In addition, there may be self-selection issues in the selection of AIPZ pilot cities in this study, such as whether a city chosen by the government as a pilot area may be influenced by the economic, industrial, or environmental conditions of the city itself. We have addressed this issue through a series of robustness tests. However, there may still be some self-selection factors in pilot cities that we have not taken into account, which may affect the robustness of the results. We will conduct a detailed analysis of the relevant policies of AIPZ, search for effective instrumental variables, and use the instrumental variable method to solve the problem of self-selection in AIPZ pilot cities.

Author Contributions: Methodology, Y.Z.; Software, C.Z.; Resources, Y.Z.; Writing—original draft, Y.Z.; Writing—review & editing, C.Z.; Funding acquisition, C.Z. All authors have read and agreed to the published version of the manuscript.

Funding: This work was supported by the Major Humanities and Social Sciences Research Projects in Zhejiang higher education institutions (No. 2024QN031). And the Achievement of the Special Project on the 'Research and Interpretation of the Spirit of the Third Plenary Session of the 20th Central Committee of the Communist Party of China and the Fifth Plenary Session of the 15th Provincial Committee of the Zhejiang Provincial Party Committee' for Social Science Planning in Zhejiang Province. And the Ningbo Philosophy and Social Science Key Research Base "Research Base on Digital Economy Innovation and Linkage with Hub Free Trade Zones".

Data Availability Statement: Data are available on request. The data are not publicly available due to the privacy and continuity of the research.

Conflicts of Interest: The authors declare no conflict of interest.

References

1. Zhou, C.B.; Sun, Z.X.; Qi, S.Z.; Li, Y.K.; Gao, H.Y. Green credit guideline and enterprise export green-sophistication. *J. Environ. Manag.* **2023**, *336*, 117648.
2. Yu, B.; Fang, D.; Pan, Y.; Jia, Y. Countries' green total-factor productivity towards a low-carbon world: The role of energy trilemma. *Energy* **2023**, *278*, 127894. [CrossRef]
3. Luo, Y.; Mensah, Z.L.; Wu, C. Environmental Regulation and Green Total Factor Productivity in China: A Perspective of Porter's and Compliance Hypothesis. *Ecol. Indic.* **2022**, *145*, 109744.
4. Taddy, M. *The Technological Elements of Artificial Intelligence*; NBER Working Paper; NBER: Cambridge, MA, USA, 2018.
5. Brynjolfsson, E.; Rock, D.; Syverson, C. *Artificial Intelligence and the Modern Productivity Paradox: A Clash of Expectations and Statistics*; Social Science Electronic Publishing: Rochester, NY, USA, 2017.
6. Agrawal, A.; Gans, J.; Goldfarb, A. *The Economics of Artificial Intelligence: An Agenda*; NBER Books: Cambridge, MA, USA, 2019.
7. Shi, B. An Explanation of the Mechanism of AI Promoting Economic High-quality Development. *Reform* **2020**, *1*, 30–38.
8. Guo, K.M. Artificial Intelligence, Structural Transformation and Labor share. *J. Manag. World* **2019**, *35*, 60–77.
9. Hu, J.; Du, C.Z. An Analysis of the Mechanism, Path, and Countermeasures of Artificial Intelligence to Promote China's Industrial Transformation and Upgrading. *Econ. Rev. J.* **2020**, *3*, 94–101.
10. Dauth, W.; Findeisen, S.; Südekum, J.; Woessner, N. *German Robots: The Impact of Industrial Robots on Workers*; Cepr Discussion Papers; IAB: Nuremberg, Germany, 2017.
11. Li, L.; Wang, X.X.; Bao, Q. The Employment Effect of Robots: Mechanism and Evidence from China. *J. Manag. World* **2021**, *37*, 104–119.
12. Huang, L.X.; Lin, Z.Y.; Wang, X.B. Industrial Robot Application and Global Value Chain Reconstruction: Based on the Perspective of Bargaining Power of Export Products. *China Ind. Econ.* **2023**, *2*, 74–92.
13. Zhao, R.Y.; Li, X.L.; Li, D.Y. Core Patents Research from the Perspective of Patent Citation: Taking the Field of Artificial Intelligence as an Example. *Inf. Stud. Theory Appl.* **2019**, *3*, 78–84.
14. Lv, Y.; Gu, W.; Bao, Q. Artificial Intelligence and Chinese Enterprises' Participation in Global Value Chains. *China Ind. Econ.* **2020**, *5*, 80–98.
15. Liu, B.; Pan, T. Research on the Impact of Artificial Intelligence on Manufacturing Value Chain Specialization. *J. Quant. Technol. Econ.* **2020**, *10*, 24–44.
16. Sun, Z.; Hou, Y.L. How Does Industrial Intelligence Reshape the Employment Structure of Chinese Labor Force. *China Ind. Econ.* **2019**, *5*, 61–79.
17. Shahzad, M.F.; Xu, S.; Naveed, W.; Nusrat, S.; Zahid, I. Investigating the impact of artificial intelligence on human resource functions in the health sector of China: A mediated moderation model. *Heliyon* **2023**, *11*, e21818. [CrossRef]
18. Wang, Y.J.; Li, Y.F. Chinese economic growth and sustainable development: Role of artificial intelligence and natural resource management. *Resour. Policy* **2023**, *85*, 103996.
19. Chao, X.J.; Zhou, W.H. Research on the Influence of Artificial Intelligence on Labor Income Share: Theoretical Interpretation and Empirical Test Based on Skill Bias. *Res. Econ. Manag.* **2021**, *2*, 82–94.
20. Cheng, C.P.; Cheng, Z. Theoretical and Empirical Research on the Mechanism of Artificial Intelligence Promoting China's Economic Growth. *Econ. Probl.* **2021**, *10*, 8–17.
21. Wang, L.H.; Qian, Y.Y.; Dong, Z.Q. Research on the Impact and Bias Effect of Artificial Intelligence Applications on Labor Wages. *Chin. J. Popul. Sci.* **2022**, *4*, 17–29.
22. Wang, L.H.; Hu, S.M.; Dong, Z.Q. Artificial Intelligence Technology, Task Attributes and Occupational Substitutable Risk: Empirical evidence from Micro-level. *J. Manag. World* **2022**, *7*, 60–79.
23. Jin, X.Y.; Zhang, W.F. Artificial intelligence and enterprise pollution reduction: Environmental effects of intelligent governance. *China Popul. Resour. Environ.* **2023**, *33*, 138–145.
24. Tian, H.N.; Zhao, L.Y.; Li, Y.F.; Wang, W. Can enterprise green technology innovation performance achieve "corner overtaking" by using artificial intelligence? Evidence from Chinese manufacturing enterprises. *Technol. Forecast. Soc. Change* **2023**, *194*, 122732.
25. Solow, R.M. Technical change and the aggregate production function. *Rev. Econ. Stat.* **1957**, *39*, 312–320.
26. Jorgenson, D.W.; Stiroh, K.J. US Economic Growth at the Industry Level. *Am. Econ. Rev.* **2000**, *90*, 161–167. [CrossRef]
27. Erkul, A.; Türköz, K. Green growth governance and total factor energy efficiency: Economic growth constraint and policy implementation in OECD countries. *Renew. Energy* **2024**, *235*, 121278.
28. Wu, H.; Hao, Y.; Ren, S. How do environmental regulation and environmental decentralization affect green total factor energy efficiency: Evidence from China. *Energy Econ.* **2020**, *91*, 104880.
29. Guo, Q.; Dong, Y.; Feng, B.; Zhang, H. Can green finance development promote total-factor energy efficiency? Empirical evidence from China based on a spatial Durbin model. *Energy Policy* **2023**, *177*, 113523. [CrossRef]
30. Gao, D.; Li, G.; Yu, J. Does digitization improve green total factor energy efficiency? Evidence from Chinese 213 cities. *Energy* **2022**, *247*, 123395.

31. Lyu, Y.; Wang, W.; Wu, Y.; Zhang, J. How Does Digital Economy Affect Green Total Factor Productivity? Evidence From China. *Sci. Total Environ.* **2023**, *857*, 159428. [PubMed]
32. Feng, C.; Ye, X.; Li, J.; Yang, J. How does artificial intelligence affect the transformation of China's green economic growth? An analysis from internal-structure perspective. *J. Environ. Manag.* **2024**, *351*, 119923. [CrossRef]
33. Ying, Y.; Cui, X.; Jin, S. Artificial intelligence and green total factor productivity: The moderating effect of slack resources. *Systems* **2023**, *11*, 356. [CrossRef]
34. Ministry of Science and Technology of the People's Republic of China. Guidelines for the Construction of the National New Generation Artificial Intelligence Innovation and Development Pilot Zone. 2019. Available online: https://www.gov.cn/zhengce/zhengceku/2019-12/03/content_5457884.htm (accessed on 1 December 2023).
35. Li, Z. Research on the Impact of Artificial Intelligence Technology on Regional Industrial Structure Transformation. *Acad. J. Bus. Manag.* **2023**, *5*, 165–169.
36. Wang, K.L.; Sun, T.T.; Xu, R.Y. The impact of artificial intelligence on total factor productivity: Empirical evidence from China's manufacturing enterprises. *Econ. Change Restruct.* **2023**, *56*, 1113–1146. [CrossRef]
37. Sestino, A.; De Mauro, A. Leveraging artificial intelligence in business: Implications, applications and methods. *Technol. Anal. Strateg. Manag.* **2022**, *34*, 16–29. [CrossRef]
38. Guo, S.; Tang, X.; Meng, T.; Chu, J.; Tang, H. Industrial Structure, R&D Staff, and Green Total Factor Productivity of China: Evidence from the Low-Carbon Pilot Cities. *Complexity* **2021**, *2021*, 6690152.
39. Jian, Z.; Bo, W. OFDI, Industrial Structure Upgrading and Green Total Factor Productivity. *Oper. Res. Manag. Sci.* **2023**, *32*, 165.
40. Lu, M.; Feng, H. Agglomeration and Emission Reduction: Empirical Study on the Impact of Urban Scale Gap on Industrial Pollution Intensity. *J. World Econ.* **2014**, *37*, 86–114.
41. Ye, Z.; Yang, J.; Zhong, N.; Tu, X.; Jia, J.; Wang, J. Tackling environmental challenges in pollution controls using artificial intelligence: A review. *Sci. Total Environ.* **2020**, *699*, 134279. [CrossRef]
42. Chen, D.K. Trade Barrier Reduction and Environmental Pollution Improvement: New Evidence from Firm-level Pollution Data in China. *Econ. Res. J.* **2020**, *55*, 98–114.
43. Zheng, S.Q.; Wan, G.H.; Sun, W.Z.; Luo, D.L. Public Appeal and Urban Environmental Governance. *J. Manag. World* **2013**, *6*, 72–84.
44. Wang, J.M.; Zhao, J. The Implementation Effect of Information Embedded Regulatory Tools on Promoting Online Green Consumption Behavior in the Digital Era: Green Purchase Scenario Simulation of and Regulatory Tools Design Experiments. *J. Manag. World* **2022**, *4*, 142–162.
45. Wu, J.; Qiu, X.; Li, Z. Green innovation and enterprise green total factor productivity at a micro level: A perspective of technical distance. *J. Clean. Prod.* **2022**, *344*, 131070. [CrossRef]
46. Zhao, X.; Nakonieczny, J.; Jabeen, F.; Shahzad, U.; Jia, W. Does green innovation induce green total factor productivity? Novel findings from Chinese city level data. *Technol. Forecast. Soc. Change* **2022**, *185*, 122021.
47. Sun, Y.; Belgacem, S.B.; Khatoon, G.; Nazir, F. Impact of environmental taxation, green innovation, economic growth, and renewable energy on green total factor productivity. *Gondwana Res.* 2023; *in press*.
48. Masoumi, Z.; van Genderen, J. Artificial intelligence for sustainable development of smart cities and urban land-use management. *Geo-Spat. Inf. Sci.* **2024**, *27*, 1212–1236. [CrossRef]
49. Zhou, D.; Hu, Y.; Sun, Q.; Xie, D. Land resource mismatch and energy efficiency: Evidence from 243 cities in China. *Energy Policy* **2023**, *183*, 113800.
50. Zhang, H.; Song, Y.; Zhang, M. How does land resource mismatch affect urban energy low-carbon transition? *Urban Clim.* **2024**, *58*, 102236.
51. Xie, R.; Yao, S.; Han, F.; Zhang, Q. Does misallocation of land resources reduce urban green total factor productivity? An analysis of city-level panel data in China. *Land Use Policy* **2022**, *122*, 106353. [CrossRef]
52. Zhou, C.B.; Qi, S.Z. Has the pilot carbon trading policy improved China's green total factor energy efficiency? *Energy Econ.* **2022**, *114*, 106268. [CrossRef]
53. Rambachan, A.; Roth, J. A more credible approach to parallel trends. *Rev. Econ. Stud.* **2023**, *90*, 2555–2591.
54. Biasi, B.; Sarsons, H. Flexible wages, bargaining, and the gender gap. *Q. J. Econ.* **2022**, *137*, 215–266. [CrossRef]
55. de Chaisemartin, C.; D'Haultfoeuille, X. Two-way fixed effects estimators with heterogeneous treatment effects. *Am. Econ. Rev.* **2020**, *110*, 2964–2996.
56. La Ferrara, E.; Duryea, S.; Chong, A.E. Soap Operas and Fertility: Evidence from Brazil. *Am. Econ. J. Appl. Econ.* **2012**, *4*, 1–31. [CrossRef]
57. Zhou, C.B.; Qi, S.Z. Can carbon emission trading policy break China's urban carbon lock-in? *J. Environ. Manag.* **2024**, *353*, 120129.
58. Li, P.; Lu, Y.; Wang, J. Does flattening government improve economic performance? Evidence from China. *J. Dev. Econ.* **2016**, *123*, 18–37. [CrossRef]

59. Wager, S.; Athey, S. Estimation and Inference of Heterogeneous Treatment Effects using Random Forests. *J. Am. Stat. Assoc.* **2018**, *113*, 1228–1242. [CrossRef]
60. Qi, S.Z.; Zhou, C.B.; Li, K.; Tang, S.Y. Influence of a pilot carbon trading policy on enterprises' low-carbon innovation in China. *Clim. Policy* **2021**, *21*, 318–336. [CrossRef]
61. An, Y. Study on the impact of land resource misallocation on urban carbon emission efficiency and its mechanism. *Urban Probl.* **2024**, *1*, 27–36.

land

MDPI

Article

Integrating Ecological Footprint into Regional Ecological Well-Being Evaluation: A Case Study of the Guanzhong Plain Urban Agglomeration, China

Xiaozheng Zheng [1], Shuo Yang [2] and Jianjun Huai [1,*]

1 College of Economics and Management, Northwest Agriculture and Forestry University, Xianyang 712100, China; zhengxz@nwafu.edu.cn

2 School of Public Administration, Xi'an University of Architecture and Technology, Xi'an 710055, China; yangshuo@xauat.edu.cn

* Correspondence: hjj@nwsuaf.edu.cn; Tel.: +86-15309253431

Abstract: This study incorporated ecological footprint (EF) consumption into a framework to assess ecological well-being. A model and implementation framework for characterizing regional net ecological well-being were then developed. Using the Guanzhong Plain Urban Agglomeration (GPUA) as a case study, land use data from 2000 to 2020 were utilized to calculate the ecosystem service value (ESV), representing the supply side of regional ecological functions. Simultaneously, the regional EF consumption was assessed as the demand side. Taking into account the level of regional economic development and the characteristics of people's living, a regional net ecological well-being evaluation model was constructed to arrive at a deficit or surplus ecological situation. The results indicated that: (1) The overall ESV of the GPUA follows a trend of initial growth followed by a decline. Woodland, grassland, and farmland are the main contributors to the total ESV, with regulating and supporting services accounting for more than 80% of the total ecosystem value. (2) EF consumption in the GPUA shows a significant upward trend, increasing by over 70% on average. The level of ecological carrying capacity has slightly increased, with the biologically productive area that can support human activities expanding to 1909.49 million hectares. Additionally, the carrying capacity of the urban agglomeration cities has tended to stabilize since 2015. (3) Since 2010, anthropogenic consumption in the GPUA has continued to exceed the regional ecological capacity, resulting in an ecological well-being deficit. The average ecological well-being compensation per hectare in the urban agglomeration increased from 35.588 CNY to 187.110 CNY. This study offers a theoretical foundation for expanding the definition and research framework of regional ecological well-being by providing a more accurate assessment of regional ecological service supply and consumption at multiple scales. It is expected that this approach will help reduce the opportunity costs associated with ecological protection, while promoting a balanced approach to economic development and ecological preservation.

Keywords: ecological well-being quota; ecosystem service value; ecological footprint; collaborative governance; Guanzhong Plain Urban Agglomeration

1. Introduction

Ecosystems provide essential services that underpin almost every aspect of human well-being, providing ecological goods and services through a variety of natural processes [1,2]. Ecological sustainability requires an increase in (appropriate) ecosystems, species and biodiversity to enhance resilience against fires, floods, heat waves, tornadoes, and other destabilizing events. At the beginning of the 21st century, the Millennium Ecosystem Assessment (MA), published by the United Nations Environment Programme (UNEP), provided a comprehensive ecological review of the complex links between human well-being and environmental functions [3]. Cities are a major contributor to global energy consumption, and cities around the world are trying to balance economic growth, environmental sustainability and energy use. Developed countries are faced with the challenge of modernizing aging infrastructure within their cities [4]. Rapid urbanization in developing countries and regions, including China, has greatly increased the demand for energy, water and land resources. As a result, ecological land, including agricultural land, is being rapidly converted for urban purposes, resulting in fragmented natural landscapes and a reduced capacity to provide ecosystem services [5]. This fragmentation impedes the full potential of ecosystems to deliver services, thus diminishing the supply of ecosystem services to urban areas and threatening their long-term sustainable development [6]. Furthermore, the depletion of ecological resources poses a significant risk to the ecological carrying capacity of these regions, creating substantial challenges to maintaining regional ecological security [7,8]. In response, China has increasingly incorporated ecosystem security and human well-being into its national strategy for building an ecological civilization. This includes the integration of resource consumption and ecological efficiency into the evaluation systems for both economic and social development.

Ecological well-being provides human societies with a range of ecological goods and services, both in quantity and quality, through complex ecological processes. The level of ecological well-being is commonly assessed by evaluating the value of ecosystem services [9]. In terms of assessment methodologies, Xie et al. established an ecosystem service value assessment system tailored to domestic ecosystems, building on the foundational work of earlier researchers [10,11]. The availability and accessibility of these ecosystem services influence their distribution and consumption, which in turn results in spatial disparities in service availability [12]. On this basis, scholars have conducted in-depth research on ecological compensation from a well-being perspective, examining the interplay between ecosystem service values and human well-being [13]. The EF transforms the ecological–economic process into a security issue by addressing the balance of supply and demand within ecologically productive spaces [14,15]. This process involves categorizing consumption demand based on land-use types and regional resource endowments, as well as assessing the offset and consumption of services provided by ecosystems [16].

Recent advances, including the three-dimensional EF method and the energy value EF model, combined ecosystem service values to enable comparative studies of ecological compensation across different regions [17]. These studies analyzed the impact of socio-economic activities on watershed ecosystems [18] and focused on the mechanisms of ecological compensation in watersheds through differential methods and spatial measurements [19]. The rest of the studies have been advancing in measuring indicators such as EF, ecological profit and loss, and ecological pressure, which are essential for evaluating the development of ecological environments [20,21].

While prior studies have explored ESV and EF independently, their integration to assess net ecological well-being remains underexplored. For instance, Deng et al. [12] emphasized the need for precise modeling of EF but did not address its impact on ecological well-being. In summary, while research on the value of ecosystem services and EF is relatively well-developed, several key areas remain underexplored: (1) Few studies have integrated ecosystem service values and EF to analyze net ecological well-being; (2) The current accounting of fossil fuel land use in EF, typically based on energy consumption per unit of GDP, is constrained by data limitations, necessitating further refinement for accuracy; (3) Most research has focused on national-level assessments, urban areas in economically developed eastern regions, or specific ecosystems such as rivers and lakes; (4) There is a noticeable gap in studies examining the relationship between EF and ecological well-being in urban agglomerations in the western regions of China.

This study made the following advances to address the shortcomings of existing studies: (1) The value of ecosystem services in the Guanzhong Plain Urban Agglomeration (GPUA) was measured with high precision using two key indicators: equivalent food value at the socio-economic level and vegetation cover index at the natural ecological level. (2) Nighttime lighting data from the DMSP-OLS satellite were used to estimate fossil energy consumption within the EF. These data were processed through a regression model calibrated with regional energy statistics to ensure data accuracy. (3) This study also assessed the current state of ecological security by investigating the ecological deficit or surplus, while synthesizing the regional level of economic development and people's standard of living. The regional GDP and Engel's coefficient were adjusted to calculate the net ecological well-being of the region, thus providing a more accurate theoretical basis for improving ecological well-being and ecological security. On this basis, targeted regional compensation programs are proposed to strengthen ecological protection and promote high-quality coordinated development of regional urban agglomerations.

This study defines ecological well-being as net ecological well-being, which refers to the value of ecosystem service provision under the constraint of EF depletion. Net ecological well-being provides a more accurate measure of the surplus or deficit in a region's ecosystems, considering the pressures exerted by human activities and social development. The primary objective of this study was to establish a comprehensive framework for assessing regional ecological well-being by integrating ESV and EF consumption [22,23]. While this study focuses on quantifying the supply–demand relationship of biophysical resources through Ecosystem Service Value (ESV) and Ecological Footprint (EF) frameworks, it is important to clarify that our assessment does not encompass the full spectrum of ecological sustainability. Specifically, aspects including biodiversity conservation, ecological resilience to climate extremes, and the cumulative impacts of emerging contaminants (e.g., microplastics, PFAS) fall beyond the scope of this quantitative analysis. Our operational definition of ecological carrying capacity herein refers specifically to the bioproductive land's capacity to support anthropogenic resource consumption, rather than pre-industrial biodiversity baselines [24,25]. Specifically, this research seeks to address the following scientific questions: (1) How can we define the ecological well-being of a region more precisely and establish relationships between ecological well-being, ESV and the EF? (2) How can the level of regional ecological well-being be calculated in relation to the level of economic and social development of the region?

The Guanzhong Plain Urban Agglomeration (GPUA) is the only urban cluster in China's national strategy explicitly designated as a pioneering area for inland ecological civilization development. Situated in northwestern China, the GPUA faces an inherently fragile ecological environment, with complex natural systems such as the Qinling Mountains, the Loess Plateau, and the Weihe River Basin. The GPUA, designated as a pilot for inland ecological civilization, faces unique challenges. These include fragile ecological environments, population pressures, and insufficient valuation of ecological products. This region is characterized by a complex and diverse geological landscape, a fragile ecological environment, and unique developmental issues, making it an ideal context for examining net ecological well-being in relation to ecosystem service provision and EF depletion. This research is essential for assessing regional ecological security, stabilizing the value of ecological products, and promoting equitable access to a healthy ecological environment. Furthermore, it holds significant academic value for the construction of ecological security barriers in arid and semi-arid regions, not only in northwest China but also in similar regions worldwide.

2. Materials and Methods

2.1. Study Area and Data Sources

The GPUA (104°34′–112°34′ E, 33°34′–36°56′ N), centered around Xi'an, is located in the inland area of northwest China and is the second largest urban agglomeration in western China. This urban agglomeration covers an area of 107,000 km^2 and is in the warm temperate continental monsoon climate zone, bordered by the Qinling Mountains in the south and the Yellow River in the east, situated between the Qinba Mountains and the Loess Plateau. In 2020, the GPUA had a resident population of 38.87 million and a regional GDP of CNY 2.2 trillion [26].

The Guanzhong Plain Urban Agglomeration Development Plan officially delineated the planning area of the GPUA. This area includes Xi'an, Baoji, Xianyang, Tongchuan, Weinan, and Shangluo (Shangzhou District, Luonan County, Danfeng County, and Zhashui County) in Shaanxi Province; Yuncheng (except Pinglu County and Quanqu County) and Linfen (Yaodu District, Houma City, Xianfen County, Huozhou City, Quwo County, Yicheng County, Hongdong County, and Fushan County) in Shanxi Province; as well as Tianshui and Pingliang (Kongdong District, Huating County, Jingchuan County, Chongxin County, and Lingtai County) and Qingyang (Xifeng District) in Gansu Province. Because of data availability, this study uses the municipal area as the boundary, including the city-wide areas of Shangluo, Yuncheng, Linfen, Pingliang, and Qingyang. The study area is shown in Figure 1 [26].

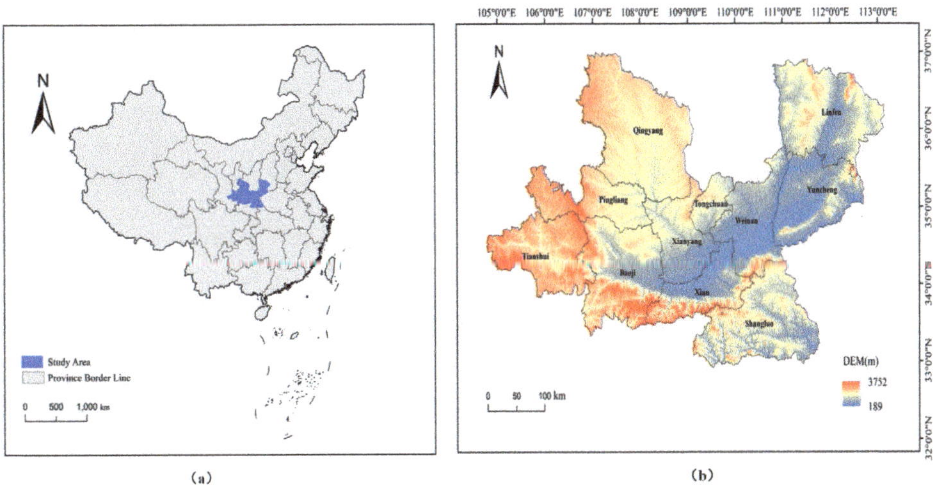

Figure 1. (**a**) Location map of the GPUA, China. (**b**) The 11 cities in the GPUA.

The data required for this study include: (1) The vector boundaries of the study area were obtained from the 1:1,000,000 National Basic Geographic Database of the National Geographic Information Resources Catalog Service System (https://www.webmap.cn/, accessed on 15 January 2023). (2) The land use data from 2000 to 2020 were obtained from the Resource Environment and Science Data Center of the Chinese Academy of Sciences (RESDC) with a spatial resolution of 30 m. The data were based on the Landsat series of U.S. Landsat satellite imagery, and were generated by human–computer interaction interpretation and manual visual interpretation (https://www.resdc.cn/, accessed on 15 January 2023). (3) Night light data (DMSP-OLS satellite and SNPP-VIIRS satellite) are from the Earth Observation Group (https://eogdata.mines.edu/products/vnl, accessed on 15 January 2023). (4) NPP data are from the MOD17A3HGF dataset of MODIS satellite called by the Google Earth Engine platform (https://www.geodata.cn, accessed on 15 January 2023). (5) Vegetation cover index data were obtained from the Resource and Environment Science Data Center of the Chinese Academy of Sciences (https://www.resdc.cn/, accessed on 15 January 2023). (6) Corresponding social and economic data and energy consumption data were obtained from the 2001–2021 Gansu Provincial Statistical Yearbook, Shaanxi Provincial Statistical Yearbook, Shanxi Provincial Statistical Yearbook, China Urban Statistical Yearbook, and statistical bulletins of cities and towns, and grain price data were obtained from the National Compendium of Agricultural Product Costs and Benefits.

2.2. Methods

2.2.1. Ecosystem Service Value

In reference to the results of Xie et al. [27], in this study, the ESV equivalent factor of one standard unit of ESV was defined as one-seventh of the economic value of natural food production per unit area of farmland per year nationwide. ESV correction and accounting were performed.

The calculation formula is as follows:

$$D = \frac{1}{7} \times \sum_{i=1}^{n} \frac{P_i G_i}{S_i} \tag{1}$$

where D is the ESV for a standard equivalent factor, P_i is the unit price of crop category i, G_i is the yield of crop category i, and S_i is the planted area of crop category i [26].

The ESV coefficient is calculated by the formula:

$$VC_j = \sum E_{ji} \times D \tag{2}$$

where VC_j denotes the coefficient of ESV of the j-th land, E_{ji} is the equivalent factor of the ESV of the i-th land use ecological service function of the j-th land, and D represents the value of a standard equivalent factor. Table 1 illustrated the ESV equivalence factors of GPUA.

In order to avoid the fluctuation of the ESV equivalent per unit area in the GPUA due to the changes in food prices in different years and the peculiarity of the correction of ESV by socio-economic factors, the Normalized Difference Vegetation Index (NDVI) was used to correct the ESV from the natural environment perspective in order to improve the accuracy of the accounting [28]. The calculation formula is as follows:

$$ESV = \sum_{i=1}^{m} \sum_{j=1}^{n} \sum_{k=1}^{o} Ee_{ij} B_{ik} C_{ik} \tag{3}$$

$$C_{ik} = \frac{NDVI_{ik}}{NDVI_i} \tag{4}$$

where Ee_{ij} is the equivalence factor for the j-th ecological service of the i-th land-use type and B_{ik} and C_{ik} are the area and NDVI coefficients of the i-th land-use type in the k-th study unit, respectively.

In Section 3.2, this study utilized the calculated ESV totals to stratify them in GIS by the natural breakpoint method and output the results.

Table 1. The ESV equivalent factors of the GPUA (CNY/hm^2·a).

Primary Type	Secondary Type	Farmland	Woodland	Grassland	Water Body	Construction Land	Unused Land
Supply service	Food production	1433.982	354.710	207.635	850.007	0.000	12.977
	Raw material production	317.942	817.564	304.964	473.668	0.000	38.932
	Water supply	25.954	423.922	168.704	7059.601	0.000	25.954
Reconciliation service	Gas regulation	1154.972	2690.608	1070.620	1732.457	0.000	142.749
	Climate regulation	603.440	8045.869	2829.031	3821.788	0.000	129.772
	Environmental purification	175.192	2340.223	934.359	5937.073	0.000	402.293
	Hydrological regulation	1940.093	5013.528	2069.865	82,061.376	0.000	272.521
Support service	Soil conservation	674.815	3274.582	1304.209	2102.308	0.000	168.704
	Nutrient maintenance	201.147	250.893	103.818	162.215	0.000	12.977
	Biodiversity	220.613	2980.432	1187.415	6761.125	0.000	155.726
Cultural service	Aesthetic landscape	97.329	1306.372	525.577	4295.456	0.000	64.886

2.2.2. EF Supply and Demand

This study adopted the ecological footprint concept proposed by William E Rees, following the accounting methodology refined by Wackemagel et al. [29]. Consumption demand is carefully divided into human consumption of natural biological and ecological resources based on land use type and regional resource endowment, and the resulting offset and consumption of services provided by ecosystems. Based on the results of existing academic research, this study used ESV as the supply side of ecological well-being and EF as the consumption side of ecological well-being, that is, the demand and depletion of ecology as manifested by human social behaviors [30,31]. Fossil energy carbon emissions, like biological resource accounts, are an important component of ecological footprint accounts, while fossil energy carbon emissions most directly affect the drivers of terrestrial ecosystem cycles [32,33].

Based on existing academic studies, in this study, in the acquisition of biologically productive land yield, based on the natural resource endowment and natural geographic and climatic conditions of the GPUA, and based on the existing research results of predecessors, the screening of biological resource types of farmland, woodland, grassland and water body was carried out in a conditional manner with 17 indicators plus 8 indicators of energy consumption, for a total of 25 indicators, and precise calculations of the required ecological capital were made in order to provide a more comprehensive and objective assessment of the state of the region's ecosystems [34]. Specific indicators were summarized in Table 2.

In this study, the equilibrium factors of farmland, woodland, grassland and water body were measured based on the NPP data in 2000, 2005, 2010, 2015 and 2020, and the EF equilibrium factor table was derived. The yield factors used in this study were taken from the Global Ecological Footprint Network [35]. Based on existing academic studies, the EF calculation methods and formulas were selected as follows:

Formula for calculating the EF model:

$$EF = N \times ef = N \cdot \sum_{i=1}^{n} (aa_i \cdot r_j) = N \cdot \sum_{i=1}^{n} (C_i / P_i) \cdot r_j \quad (5)$$

Table 2. Indicators for EF accounting.

Resource Account	Account Name	Productive Land-Use Type	Selected Indicators
Biological Resources Accounts	Agricultural Products Account	Farmland	Grain, Cotton, Oilseeds, Tobacco, Vegetables, Pork
	Forest Products Account	Woodland	Apple, Pear, Grape, Peach, Apricot
	Grassland Products Account	Grassland	Beef, Lamb, Goat's wool, Poultry eggs, Honey
	Water Products Account	Water body	Aquatic product
Energy Consumption Accounts	Energy Consumption Account	Construction Land	Raw coal, Coke, Petrol, Paraffin, Diesel, Fuel oil, Natural gas, Electricity

Following the Global Footprint Network's methodology, the EC in this study is defined as the biologically productive area available to provide renewable resources and absorb waste under current management practices. This should be distinguished from the broader concept of ecological sustainability that requires maintaining evolutionary potential and ecosystem integrity. The human carrying capacity component focuses on the balance between resource provisioning services and socioeconomic demands, not encompassing demographic projections or technological disruption scenarios [36].

Formula for calculating EC:

$$EC = N \times ec = N \times \sum_{j=1}^{n} (a_j \times r_j \times y_j) \tag{6}$$

where EF is the ef produced by the total population of the region (hm^2); EC is the ec of the total population of the region (hm^2); N is the total population in the study area; aa_i is the converted bio-productive area per capita of the i-th consumption item; C_i and P_i correspond to the per capita consumption and average production capacity of the first consumption item, respectively; a_j is the per capita ecologically productive area of the j-th land-use type; and r_j and y_j are the equilibrium factor and yield factor of the j-th land-use type, respectively.

Carbon emissions inverted with nighttime lighting data were converted to energy consumption in units of million tons of standard coal, as well as the global average energy footprint and conversion factor, which converted the heat consumed by the municipality's energy consumption to fossil energy land area, and then the EF and EC models were used to derive the EF and EC corresponding to its energy account [32].

2.2.3. Measurement of Regional Ecological Safety Factor and Ecological Well-Being

The regional ecological security index is an important indicator for measuring regional ecological environmental problems and carrying out ecological environmental protection. Based on the EF and EC model, it can be used as a consideration for regional ecological security, and is the basis for analyzing whether regional ecological well-being is in a deficit or surplus state.

The ecological security coefficient thus determined is:

$$ES = \frac{EF}{EC} \tag{7}$$

where ES is the regional ecological security index. When $ES > 1$, the regional ecological security belonging to the ecological consumption is beyond the carrying range, that is, the ecological well-being is in a deficit state; and when $ES < 1$, the regional ecological security belonging to the ecological consumption is in the ecological carrying range, that is, the ecological well-being is in a surplus state.

Considering the constraints of multiple factors, such as regional economic, social and ecosystem functions, for measuring the amount of ecological well-being, this study constructed a model for accounting for ecological well-being by using the Pearl growth curve, the regional GDP and the Engel's coefficient, constraining it within the framework of the local economy and standard of living. The resulting regional ecological well-being measure more accurately represents the level of ecological well-being in the region.

$$E_i = R_i ESV_i |EC - EF| \tag{8}$$

$$R_i = \frac{e^{\varepsilon} GDP_i}{GDP(e^{\varepsilon} + 1)} \tag{9}$$

where E_i is the total regional ecological well-being, R_i is the regional development coefficient adjusted to take into account the regional level of economic development, ESV_i is the value per unit area of the regional ecosystem services, EC is the regional ecological carrying capacity, EF is the regional EF, e^{ε} is a function of the natural logarithm of the base, ε is the regional Engel's coefficient, GDP_i is the total ecological value of the area i within the region, and GDP is the gross regional product.

3. Results

3.1. Land Use Change Analysis

Figure 2 and Table 3 illustrate the land-use types of the GPUA from 2000 to 2020. According to the land use classification standards, the GPUA is predominantly composed of farmland, grassland, and woodland, which together accounted for approximately 94.7% of the total land area in 2020. Specifically, farmland made up about 39.9%, grassland constituted around 33%, and woodland covered approximately 21.8% of the region's total land area.

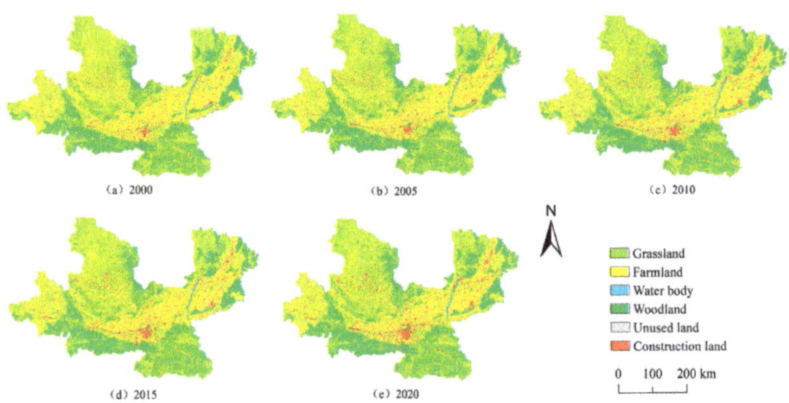

Figure 2. Land use changes in the GPUA over the study period.

Table 3. Changes in land use structure of the GPUA.

Periods	Types	2020						
		Farmland	Woodland	Grassland	Water Body	Construction Land	Unused Land	Total (km^2)
2000	Farmland	61,607.137	805.749	3724.272	231.531	2212.084	24.307	68,605.081
	Woodland	369.481	33,318.434	662.822	14.221	55.792	10.318	34,431.068
	Grassland	2206.467	1166.475	48,937.062	51.787	139.740	21.090	52,522.622
	Water body	146.864	10.015	46.891	1232.455	36.638	13.987	1486.850
	Construction land	389.774	16.014	31.058	8.593	4387.668	0.537	4833.644
	Unused land	12.811	5.150	13.498	15.614	4.164	120.052	171.289
	Total	64,732.535	35,321.838	53,415.602	1554.201	6836.086	190.292	162,050.554

Geographically, the GPUA benefits from a diverse landscape. The northern part of the urban agglomeration features loess hilly and ravine areas with relatively high vegetation cover, as well as rich grassland and forest resources. The central region consists of the Guanzhong and Fenwei Plains, which are adjacent to the Weihe River system, making them highly suitable for agricultural development. The southern area is dominated by the Qinling Mountains, known for their ecological diversity. However, the ecological quality along the Weihe River is relatively poor, with a further reduction in biodiversity due to large-scale agricultural activities. The Guanzhong Plain Urban Agglomeration Development Plan has identified several ecological corridors, which play a critical role in maintaining the ecological barrier between the loess hills and gullies and the Qinba Mountain [7].

3.2. *Multidimensional Analysis of ESV*

To estimate the ESV of the GPUA, data on the production of major grain crops, sown area, and grain purchase prices from 2000 to 2020 were used. This allowed for the calculation of the ESV coefficient specific to the GPUA. Additionally, the natural ecosystem was corrected using the Normalized Difference Vegetation Index (NDVI), which provides a more accurate reflection of the natural environment's influence on the ecosystem. The corresponding service-type equivalent factors for ESV were determined by processing 1 km × 1 km resolution NDVI data using ArcMap (10.8.1) software.

The results can be seen from the Table 4 that the total ESV of the GPUA exhibited fluctuating trends from 2000 to 2020, with an overall pattern of increase followed by a decrease. The peak value of ESV reached 180,472.5 million CNY in 2010, after which it declined from 169,467.2 million CNY to 167,298.3 million CNY by 2020, representing a decrease of 1.28%. In terms of ESV contributions by land-use types, woodland, grassland, and farmland were the primary contributors to the total ESV. Construction land, in contrast, was the main land-use type experiencing growth, with an increase of 2002.44 km^2. Water bodies, though covering a smaller area (67.351 km^2), saw a significant ESV increase of 593.2 million CNY, representing the highest percentage increase (4.48%) among land-use types. This suggests that water bodies have a high unit ESV and provide considerable provisioning services, highlighting their critical ecological importance. The fluctuations in total ESV during the study period can be attributed to land use conversions between categories [26].

Table 4. Total ESV for different land-use types in the GPUA (billion CNY·hm^2).

Land-Use Type	ESV (Billion·hm^2)					2000–2010 Amount of Change/Rate of Change	2010–2020 Amount of Change/Rate of Change	2000–2020 Amount of Change/Rate of Change
	2000	2005	2010	2015	2020			
Farmland	352.406	352.314	383.827	349.791	342.349	31.421 /8.92%	−41.479 /−10.81%	−10.057 /−2.85%
Woodland	784.878	798.818	828.715	803.878	790.139	43.837 /5.59%	−38.577 /−4.65%	5.260 /0.67%
Grassland	424.903	438.309	452.894	436.554	402.059	27.991 /6.59%	−50.835 /−11.22%	−22.844 /−5.38%
Water body	132.299	144.038	139.121	128.923	138.231	6.821 /5.16%	−0.890 /−0.64%	5.932 /4.48%
Construction land	0.000	0.000	0.000	0.000	0.000	0	0	0
Unused land	0.186	0.175	0.168	0.146	0.206	−0.019 /−9.96%	0.038 /22.65%	0.019 /10.43%
Total	1694.672	1733.654	1804.725	1719.292	1672.983	110.052 /6.49%	−131.742 /7.30%	−21.690 /−1.28%

The spatial and temporal distribution of total ESV is illustrated in Figure 3, with the total ESV classified into four intervals: <90, 90–170, 170–250, and >250 million CNY. The results indicate an overall upward trend in the total ESV of the GPUA. Notably, the number of cities within the <9 billion CNY range has decreased, while approximately half of the cities in the agglomeration now fall within the 9–17 billion CNY range. This change is primarily attributed to the expansion of ecological land areas, such as woodland and grassland, within the region. The total ESV exhibits a distinct spatial distribution pattern of 'high in the surrounding areas and low in the center,' and 'high in the east and low in the west,' which mirrors the overall land use structure.

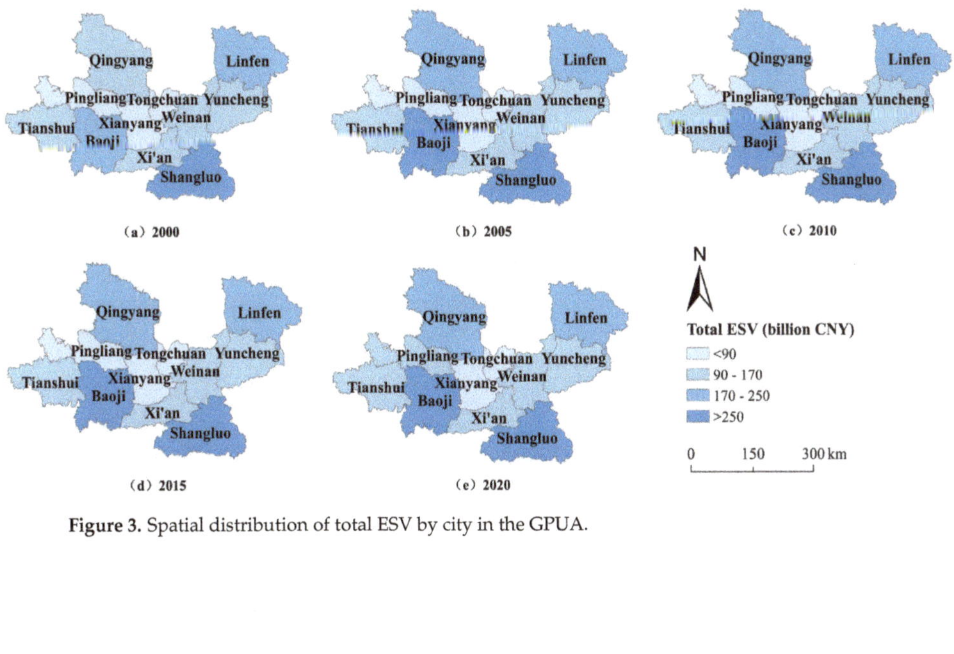

Figure 3. Spatial distribution of total ESV by city in the GPUA.

3.3. EF and EC Accounting and Spatial and Temporal Pattern Release Analyses

In this study, carbon emission data for each city were derived from nighttime lighting data and fitted, providing reasonable estimates of the carbon emission levels for each city. Based on this, the ecosystem resource consumption in both the biological resource and fossil energy sub-accounts was measured. Figure 4 depicted the EF consumption of each city in the GPUA.

Figure 4. EF of cities in the GPUA (10^4 hm^2).

The results revealed that only Shangluo, Tianshui, and Pingliang showed a slight downward trend in their EF in 2020, although the total EF remained significantly higher than at the start of the study period. The accelerated urbanization in the GPUA has led to increased land development, with construction land becoming the primary site for fossil fuel use, directly contributing to changes in carbon emissions and sequestration, and subsequently affecting the capacity of ecosystem services.

To better illustrate the total EF consumption in the GPUA and its spatial distribution over the study period, Figure 5 presents the changes in total EF across different cities, categorized into four intervals: 0–120, 120–240, 240–360, and >360. These results are detailed in Figure 5.

After 2010, the EF of the entire urban agglomeration increased sharply (Tongchuan is excluded from the regional trend analysis due to its size and population). The EF rose in the third and fourth intervals, 240–360 and >360, respectively. The distribution of total consumption shifted from a pattern of "west–central–east" to "center–surround," reflecting a clear geographical redistribution of ecological resource consumption across the entire urban agglomeration. This pattern provides a comprehensive view of the changing ecological resource demands within the region.

From the calculations in Table 5, it can be analyzed that the level of ecological carrying capacity of the region has slightly increased in2020, and the ecological carrying capacity from 2000 to 2020 has been calculated to be: 2402.520 $\times$ 10^4 hm^2, 2498.137 $\times$ 10^4 hm^2, 2513.363 $\times$ 10^4 hm^2, 2589.526 $\times$ 10^4 hm^2 and 2593.469 $\times$ 10^4 hm^2, respectively, an increase of 1909.49 million hectares. These fluctuations can be attributed to changes in land-use types and the long-term effects of the region's ecological protection policies, both of which have contributed to the variations in the region's ecological carrying capacity.

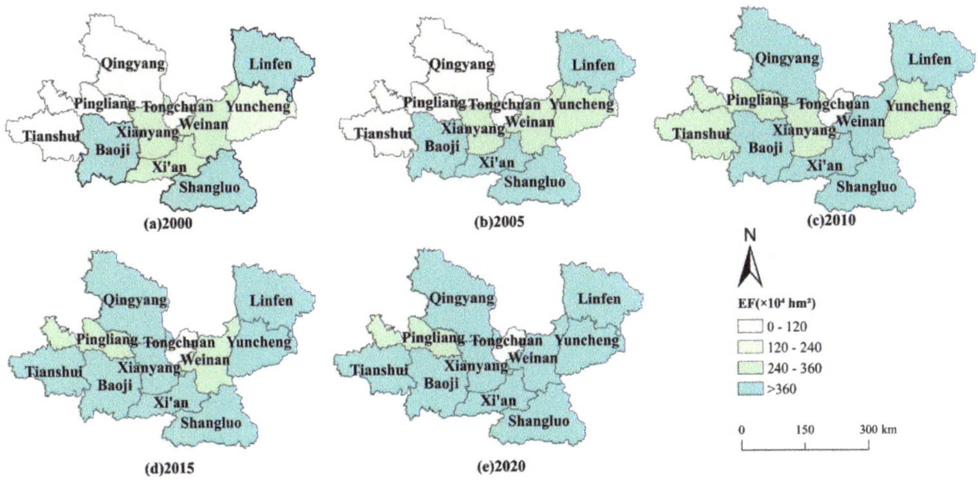

Figure 5. Spatial and temporal distribution of the total EF of the GPUA.

Table 5. EC of the total account of biological resources in the GPUA (10^4 hm^2).

Year	Baoji	Linfen	Pingliang	Qingyang	Shangluo	Tian Shui	Tong Chuan	Weinan	Xi'an	Xianyang	Yuncheng
2000	261.182	298.029	163.145	373.304	268.804	207.708	57.717	161.656	209.254	166.026	235.697
2005	271.106	309.606	169.176	388.532	278.608	215.178	59.931	169.192	218.312	173.188	245.308
2010	273.300	313.426	166.902	388.697	280.102	216.046	60.246	172.566	220.230	175.237	246.612
2015	281.372	322.623	171.821	399.625	288.107	222.474	62.023	178.932	227.581	180.802	254.167
2020	280.404	323.809	171.697	398.668	286.958	223.078	61.821	179.218	229.303	181.983	256.529

From the calculations in Figure 6, it can be analyzed that Pingliang, Xianyang and Xi'an gradually rose from the second range of 70–170 to the third range by 2020, and the distribution of ecological carrying capacity changed from the 'high north–low middle' and 'high east–low west' patterns in 2000 to a better overall distribution pattern. The distribution of ecological carrying capacity changed from the pattern of 'high in the north–low in the middle' and 'high in the east–low in the west' in 2000 to a better overall distribution pattern, and the carrying level of each city in the urban agglomeration tended to be stable after 2015. The continuous expansion of construction land has caused the areas of woodland, grassland and water bodies around urban ecosystems to shrink, impeding the functions of nourishment and regulation, reducing the ability to prevent and control environmental risks, and exacerbating the risks and pressures on regional ecological and environmental security.

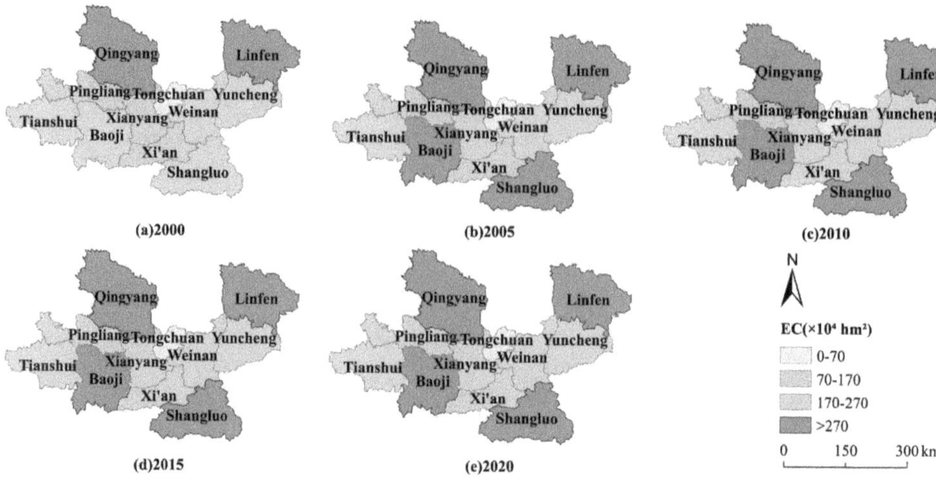

Figure 6. Spatial and temporal distribution of the total EC of the GPUA.

3.4. Formatting of Mathematical Components

The comparative measurement of the total EF and EC of the bioresource account was used to derive the ecological security index as well as the ecological capacity of each city for the years 2000, 2005, 2010, 2015 and 2020, and the results are shown in Table 6 below.

Table 6. Ecological security index and zoning of cities in the GPUA.

Year	Baoji	Linfen	Pingliang	Qingyang	Shangluo	Tian Shui	Tong Chuan	Weinan	Xi'an	Xianyang	Yuncheng
2000	1.408 deficit	1.510 deficit	0.238 surplus	0.110 surplus	1.596 deficit	0.220 surplus	0.905 surplus	1.260 deficit	1.910 deficit	1.647 deficit	0.950 surplus
2005	1.489 deficit	1.760 deficit	0.289 surplus	0.140 surplus	1.620 deficit	0.250 surplus	0.975 surplus	1.388 deficit	2.436 deficit	1.856 deficit	1.082 deficit
2010	1.564 deficit	1.786 deficit	1.807 deficit	1.966 deficit	1.645 deficit	1.655 deficit	1.148 deficit	1.636 deficit	2.656 deficit	2.045 deficit	1.257 deficit
2015	1.836 deficit	1.838 deficit	1.813 deficit	2.123 deficit	1.654 deficit	1.653 deficit	1.441 deficit	1.573 deficit	2.918 deficit	2.076 deficit	1.418 deficit
2020	2.095 deficit	2.316 deficit	1.822 deficit	2.170 deficit	1.646 deficit	1.644 deficit	1.861 deficit	1.914 deficit	4.318 deficit	2.741 deficit	2.124 deficit

From the analysis of the results shown in Table 7, it can be seen that the GPUA is generally in a state where ecological consumption is greater than the ecological carrying range, and is in a state of ecological capacity deficit. Consistent with the characteristics presented by the ecological safety coefficient, Pingliang, Qingyang and Tianshui in Gansu Province were in a surplus state of ecological well-being amount in 2000–2005; Tongchuan in Shaanxi Province was in a surplus state in 2000–2005; and Yuncheng in Shanxi Province was in a surplus state in 2000. In the remaining years, all cities were in deficit, with the overall deficit widening to 1378.9 million CNY in 2020, or 1364.914 CNY per hectare, in Xi'an.

Table 7. Total amount of ecological well-being of cities in the GPUA (billion CNY·hm^2).

Year	Baoji	Linfen	Pingliang	Qingyang	Shangluo	Tian Shui	Tong Chuan	Weinan	Xi'an	Xianyan	Yuncheng
2000	−0.763	−0.830	0.438	0.927	−0.333	0.749	0.007	−0.280	−3.277	−0.798	0.061
2005	−0.879	−1.685	0.429	1.327	−0.325	0.765	0.002	−0.382	−4.980	−0.879	−0.123
2010	−0.891	−1.427	−0.480	−1.762	−0.332	−0.649	−0.009	−0.586	−5.487	−1.011	−0.325
2015	−1.378	−1.518	−0.436	−2.127	−0.413	−0.727	−0.026	−0.546	−6.950	−1.166	−0.579
2020	−1.763	−2.061	−0.435	−1.967	−0.362	−0.619	−0.050	−0.814	−13.789	−1.479	−1.452

It can be observed from Figure 7 that the more obvious of these showed that all cities are in a declining trend and basically reached their lowest values in 2020. Baoji, Shangluo, Xianyang, Tianshui, Pingliang and Qingyang had a small rebound, but the overall trend was still declining. It can be seen that the ecological well-being of the GPUA presents a distribution trend of high in the surrounding area and low in the center.

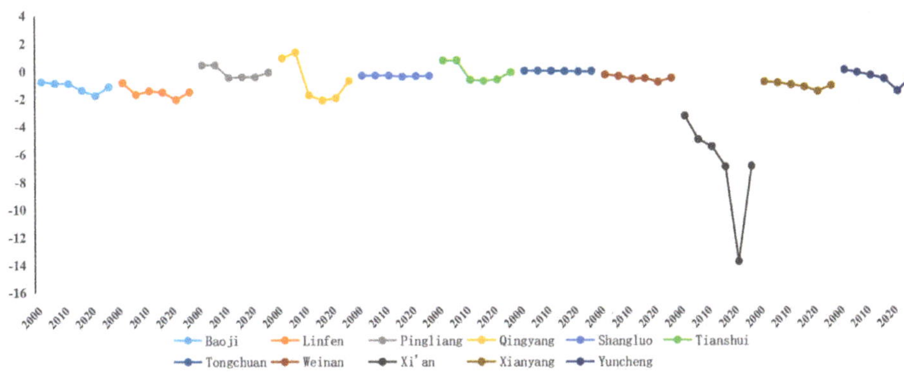

Figure 7. Eco-well-being amount and change trend of each city in the GPUA.

The southern part of Gansu Province, to which the city in question belongs, is rich in biological resources, with a relatively high proportion of grassland and woodland, and its ecosystems provide greater provisioning and regulating functions and values than other types of land. Moreover, in recent years, the local government has focused on ecological environmental protection and 'three lines and one single' ecological zoning control, with a total of 56 priority protection, key control and general control units in Tianshui, a total of 61 in Pingliang, and a total of 72 in Qingyang, and the control of key areas in the southern Qinba Mountains and the Gannan Plateau has achieved certain results. Baoji and Shangluo have actively implemented the key tasks of *the Regulations on the Protection of the Ecological Environment of the Qinling Mountains in Shaanxi Province*, as well as paying attention to the environmental protection of mining geology and land reclamation, and have implemented strong initiatives to do a good job of ecological environmental protection.

4. Discussion

4.1. Spatial and Temporal Heterogeneity in the Amount of the ESV

From 2000 to 2020, the total ESV demonstrated fluctuating growth, with two periods of significant change: from 2000 to 2010, and from 2015 to 2020. This bimodal growth pattern aligns with recent findings in China's Yellow River Basin, where policy implementation lags and ecological project phasing were identified as key drivers of non-linear ESV trajectories [37]. This temporal pattern aligns with recent studies on China's ecological restoration policies, which identified similar phased ESV growth driven by the "Ecological

Civilization" strategy and the 13th Five-Year Plan [38]. The distribution of ESV values in the southern, western, and northwestern edges of the study area—specifically in the Qinba Mountains and along the Yellow River—can be attributed to the positive effects of forest cover and water resource protection on ESV intensity [39]. Recent remote sensing analyses demonstrated that mountainous regions with >60% forest cover (similar to the Qinba Mountains) exhibited 2–3 times higher ESV density compared to lowland urban areas, supporting our spatial observations [38]. Recent global analyses (e.g., in the Himalayas and Andes) confirmed that mountainous regions with high forest cover and water retention capacity consistently exhibit elevated ESV heterogeneity, driven by topographic gradients and conservation policies [40]. The rapid growth of the total ESV in 2005 is largely due to the expansion of woodland area, marking the largest increase during the study period. This finding contrasts with previous studies that emphasized cropland conversion as the primary driver of ESV [41], and at the same time, this finding contrasted with studies in eastern China (e.g., Yangtze River Delta), where urbanization dominated ESV declines during the same period, highlighting the unique role of the GPUA as an ecological transition zone, but was consistent with recent work showing the dominance of afforestation in China's post-2000 ecological restoration programs [42]. Additionally, the retreat of farmland, grassland projects, and erosion control efforts provided a strong foundation for the restoration of ecosystem services. Comparative analysis revealed that the GPUA's farmland conversion rate (8.7% during 2000–2010) exceeded the national average of 5.3% reported by Bryan et al. [37], but matched recent regional data from the Loess Plateau (8.2–9.1%) [43], suggesting localized policy prioritization. A 2022 meta-analysis of 126 global cases revealed that farmland conversion to forests in semi-arid regions (like the GPUA) increased ESV by 18–24%, exceeding the global average (12–15%), likely due to the synergistic effects of reduced soil erosion and enhanced carbon sequestration [44].

Efforts to build green barriers, such as the windbreak and sand trap forest belt, the Loess Plateau Soil and Water Conservation Ecological Zone, the Weihe River Ecological Landscape Belt, and the Qinba Mountain Biodiversity Ecological Functional Zone, have been crucial in this process [45]. These multi-scale interventions mirror the "ecological corridor" approach advocated in the 2020 National Ecological Security Strategy, but with innovations in integrating desert–loess transition management—a critical gap identified in previous studies [46]. These initiatives have strengthened the ecological security of the region and contributed to the overall increase in ESV.

The provinces and prefectural-level cities within the GPUA urban agglomeration should independently compile records and accounts detailing the functional value provided by ecosystem services, ecological product value, environmental pollution, EF consumption, ecological remediation, and carbon emission rights trading. By quantifying responsible parties and compensation standards, the problem of environmental externalities can be effectively addressed. The distribution of compensation funds and the corresponding measures will be executed accordingly. Different provinces and cities may establish separate agencies to manage ecological well-being enhancement and compensation. These agencies will be tasked with addressing and coordinating the implementation of policies, as well as resolving conflicts and disputes among the primary stakeholders within the urban agglomeration during the ecological compensation process. Under the guidance of higher-level organizations, these agencies will decompose tasks and regulate implementation actions based on account preparation and the principles of ecological well-being enhancement.

The GPUA should focus on promoting systematic management projects for mountains, rivers, forests, fields, lakes, grasslands and deserts. This approach will improve carbon sequestration in urban green landscapes and natural ecological areas, as well as enhance the regulatory functions of climate regulation, air purification, and the water conservation capabilities of watersheds, lakes, and wetlands. These efforts will contribute to the rapid development of the "Million Acres of Forests" and "Million Acres of Wetlands" projects in the GPUA. The successful construction of these projects in the Guanzhong region will help reduce the opportunity cost of ecological protection. Additionally, this approach will facilitate a coordinated division of labor between economic development and ecological protection, ensuring that ecological compensation efforts align with the region's long-term sustainability goals. This spatial synergistic governance model can reduce the opportunity cost of ecological protection and achieve Pareto improvements in economic–ecological efficiency through differentiated division of labor.

Market mechanisms can be used to complement and improve carbon management tools and scientifically control the total amount of carbon emissions. On this basis, the initial allocation system of carbon emission rights will be used to carry out market-based environmental rights trading so as to achieve the common governance effect of diversified ecological governance and compensation for loss credits [47,48]. In the future process of urban ecosystem protection and problem management in the GPUA, in addition to satisfying basic provisioning, regulating and supporting services, it is important to focus on aesthetic creation and landscape gardening to enhance the aesthetic landscape value of urban green space, taking into account both functionality and artistry [49,50]. In 2018, China promulgated the Guanzhong Plain Urban Agglomeration Development Plan, which emphasizes ecological environmental protection as both a task and a prerequisite for the development of urban clusters. This policy provides crucial support for the rapid growth of ecosystem service values and the implementation of ecological protection and restoration projects from 2015 to 2020.

4.2. Rationality and Necessity of Calculating Ecological Well-Being Based on Ecological Footprint Depletion

Ecosystems in western China have a significant influence on sustainable development and human well-being not only in eastern China but also across broader regions of Asia [51–53]. From 2000 to 2020, the overall EF consumption in the study area exhibited a marked upward trend, primarily driven by fossil energy consumption, in line with urban and economic development. This aligns with global patterns reported by Wiedmann et al. (2020), who identified fossil energy as the dominant driver of EF growth in developing economies, particularly in post industrialization regions [54]. By inverting carbon emissions using nighttime lighting data, it was found that the carbon emissions from construction land exceeded the net carbon emissions for the region [50]. Recent advances in spatial carbon accounting, such as the "light-based emission mapping" method validated by Wan et al. (2024) for Chinese cities, demonstrated that this approach can effectively capture high-resolution spatial patterns of urban CO2 emissions [55].

Human activities, particularly the intensive exploitation of land, directly impact the supply capacity of ecosystem services and functions. Taking the GPUA as a case study, the ecological footprint has risen substantially, with geographic changes in consumption expanding from the urban center to surrounding areas. Similar spatial diffusion patterns were observed in the Beijing–Tianjin–Hebei region by Chen et al. (2024), who attributed this "ecological spillover" to urban agglomeration effects and interregional resource dependencies [56]. However, since 2010, the ecological carrying capacity of the GPUA has largely stabilized, although the ecological footprint continues to exert a significant negative influence on the region's ecological security. The decoupling of EF growth and carrying capacity stabilization reflects the "ecological redline" policy effects observed in the Beijing–Tianjin–Hebei region, though the GPUA achieved this transition 5–7 years earlier due to stricter land-use control [57].

The average compensation rate per hectare for ecological well-being in the GPUA increased substantially, from 35.59 CNY to 187.11 CNY. The correlation between the compensation price per hectare and the changes in the ecological well-being quota for each city further highlights the alignment of these two trends [32].This compensation mechanism echoes the "payment for ecosystem services" (PES) framework proposed by Salzman et al. (2018), but innovates by integrating industrial restructuring into compensation criteria—a critical advancement for urban agglomerations [58].

Xi'an, as the core of the GPUA, has accelerated urban construction and social development, driving industrial transformation and upgrading. The secondary and tertiary sectors now account for approximately 90% of the region's GDP, which aligns with the findings of previous studies [2,59], thereby improving the accuracy and robustness of the accounting method used in this study. Notably, this industrial structure shift contrasts with the "tertiary-led decarbonization" model observed in the Pearl River Delta, where service industries contributed 75% of GDP but only 40% of EF reduction, suggesting the GPUA's manufacturing-focused transition may yield higher ecological efficiency [60]. However, a mismatch exists between the economic space and ecological resources in the GPUA. This rapid industrialization and urbanization process have intensified this imbalance. The GPUA with Xi'an as the core has accelerated the process of industrial transformation and upgrading in the process of rapid urban construction and social development, and the output value of the secondary and tertiary industries accounts for about 90% of the regional GDP, which is basically consistent with the results of the previous relevant studies [1]. For the achievement of the rationality of regional ecological industrial division of labor and the improvement of the efficiency of ecological well-being, the GPUA should form an effective and differentiated division of labor in the ecological protection trade-offs based on the differences in the geographic locations and resources of different cities. Addressing the challenge of expanding urban development while preserving ecological integrity remains a critical issue for further exploration and policy formulation.

To reduce ecological resource consumption, particularly fossil energy use, it is recommended to establish regional carbon emission caps. For instance, Xi'an and Xianyang could implement stricter building energy efficiency codes, while Linfen and Yuncheng could prioritize the deployment of clean coal technologies. Strengthening territorial management and addressing the negative externalities of air pollutants should be prioritized. To this end, the implementation of a Joint Prevention and Control Mechanism is recommended to promote regional synergy in air quality management. This strategy should serve as a key criterion for assessing ecosystem protection and the enhancement of well-being within the region.

At the same time, innovation platforms such as the National Engineering Research Center and the Western China Innovation Port have helped to overcome the constraints of the traditional industrial chain, especially in the key areas of aerospace power, materials and research and development. The development of core technologies could be strengthened by overcoming technological bottlenecks. Traditional heavy industrial cities such as Linfen and Yuncheng should focus on exploiting their advantages in energy and chemical resources, while solving key technological difficulties in energy-intensive industries such as coal-fired power generation and coal–oil co-processing. At the same time, Xianyang should prioritize its role as the core of the Xi'an metropolitan area and play a demonstration role in developing the aviation industry, etc. Through these efforts, a differentiated division of labor can be formed based on the united relationship between ecological resources, functional product supply, and industrial development, allowing the region to capitalize on its comparative advantages within the ecological protection trade-off process. This industrial restructuring based on comparative advantages can make ecological protection an endogenous variable rather than an external constraint for high-quality development.

4.3. Research Limitations and Prospects

Three critical limitations frame the interpretation of our findings: First, the ESV-EF framework prioritizes quantifiable provisioning services over biodiversity indicators and ecological resilience. Second, the exclusion of cumulative stressors like chemical pollutants and microplastics may overestimate sustainable yields. Third, our carrying capacity calculations assume static technological conditions, whereas disruptive innovations could alter future resource equations. A significant limitation of this study is the exclusion of soil quality and precipitation variability in the correction of ESV equivalent factors. These factors are critical in regions like the GPUA, where soil erosion and water availability significantly impact ecosystem services. In the process of EF accounting, biological resources and energy consumption were considered, but carbon sink capacity was not taken into account when measuring EC. Consequently, after considering the constraints of ecosystem service provision and human resource consumption and carrying level, it is important to explore the regional ecological well-being amount and well-being enhancement compensation.

In future studies, the methodology can be further refined by incorporating more precise and accurate indicator data. Subsequent research should integrate molecular biodiversity monitoring (e.g., eDNA metabarcoding) with dynamic modeling of contaminant pathways to develop next-generation sustainability metrics. Future research should integrate high-resolution soil maps and climate data to enhance the robustness of ESV corrections. By combining land use data, researchers can scale up the measurement of regional ecological well-being and explore additional natural and socio-economic driving factors.

5. Conclusions

This study constructed a regional ecological well-being accounting and realization model focusing on GPUA. Based on land use data from 2000 to 2020, the model aimed to elucidate the relationship between the value of services provided by natural ecosystems and the consumption of urban ecosystems by human activities. The ESV and EF models were used to quantify the regional ecological service supply and human consumption and analyze their spatial distribution characteristics. By measuring ecological capacity, a regional security index was calculated. In addition, the results of ecological well-being surplus or deficit were constrained by GDP and Engel's coefficient. The "net ecological well-being" indicator in this study represented relative progress in resource accounting rather than absolute sustainability. This has improved the accuracy of assessing regional net ecological well-being and enhanced the robustness and completeness of the results of previous studies.

(1) From 2000 to 2020, the total ESV in the GPUA exhibited fluctuating changes, with a general trend of increase followed by a decline, peaking in 2010. The distribution of ESV value across the cities within the GPUA is imbalanced. High-value areas are predominantly located in regions with high altitude and slope, such as the Yellow River and Weihe River Basin areas, which are rich in ecological resources and feature more intact ecosystems. These areas also exhibit significant positive ecological effects. Water bodies, which provide high ESV intensity per unit area, contribute substantially to the overall ESV. They play a critical role in ecosystem services, including both provisioning and regulating functions, underscoring their importance in the region's ecological structure.

(2) Overall, the GPUA has experienced a substantial increase in its total consumption of ecological resources, with fossil energy consumption comprising more than one-third of the

total consumption of bioproduction in each city. This consumption exceeds the ecological carrying capacity by five to six times. The development of many cities, such as Xi'an and Yuncheng, still relies more on regional mineral resources and traditional high-energy-consuming industries. The region's EC capacity has shown a modest increase, with an additional 1.91 million hectares of biologically productive area capable of sustaining human activities. The distribution of EC has changed from the 2000 pattern of 'high in the north—low in the center' and 'high in the east—low in the west' to a better overall distribution pattern, and the carrying capacity levels of cities within the urban agglomeration have tended to stabilize after 2015. After 2015, the carrying capacity of each city in the urban agglomeration tended to stabilize.

(3) After 2010, the GPUA experienced a significant ecological deficit, where human consumption of ecological resources surpassed the regional ecological carrying capacity. The overall ecological well-being showed an increasing deficit, reaching its lowest value in 2020. Notably, Xi'an has a much larger ecological well-being deficit than other cities in the region, rising from −324.45 CNY in 2000 to −1364.91 CNY in 2020. The ecological well-being deficit in the GPUA follows a distinct spatial pattern: higher in the western regions, lower in the central and eastern parts, and greater in the surrounding areas compared to the urban center.

Author Contributions: X.Z.: conceptualization, conceptualization, methodology, formal analysis, investigation, data curation, writing—original draft. S.Y.: methodology, validation, supervision, and review. J.H.: methodology, validation, supervision, review and funding acquisition. All authors have read and agreed to the published version of the manuscript.

Funding: This research was funded by the National Natural Science Foundation of China (No. 42075172) and the National Social Science Foundation of China's Western Region Project (No. 23XTJ006).

Data Availability Statement: Dataset available on request from the authors.

Acknowledgments: The authors would like to thank the anonymous reviewers for their constructive comments that helped improve this paper.

Conflicts of Interest: The authors declare no conflicts of interest.

Abbreviations

The following abbreviations are used in this manuscript:

GPUA	Guanzhong Plain urban agglomeration
EF	ecological footprint
ESV	Ecosystem service value
EC	ecological carrying capacity

References

1. Nie, X.; Li, X.J.; Lyu, C.; Su, Y.L.; Wang, H. Can Ecological Compensation Based on the Transfer of Development Rights (Tdr) Improve Ecosystem Service Value? A Multi-Scenario Simulation. *Land Use Policy* **2024**, *138*, 107024. [CrossRef]
2. Cao, Y.; Kong, L.; Zhang, L.; Ouyang, Z. The Balance between Economic Development and Ecosystem Service Value in the Process of Land Urbanization: A Case Study of China's Land Urbanization from 2000 to 2015. *Land Use Policy* **2021**, *108*, 105536.
3. Leemans, R.; De Groot, R. *Millennium Ecosystem Assessment: Ecosystems and Human Well-Being: A Framework for Assessment*; World Resources Institute: Washington, DC, USA, 2003.
4. Yatzkan, O.; Cohen, R.; Yaniv, E.; Rotem-Mindali, O. Urban Energy Transitions: A Systematic Review. *Land* **2025**, *14*, 566. [CrossRef]
5. Frota de Albuquerque Landi, F.; Di Giuseppe, A.; Gambelli, A.M.; Palliotti, A.; Nicolini, A.; Pisello, A.L.; Rossi, F. Life Cycle Assessment of an Innovative Technology against Late Frosts in Vineyard. *Sustainability* **2021**, *13*, 5562. [CrossRef]

6. Straffelini, E.; Pijl, A.; Otto, S.; Marchesini, E.; Pitacco, A.; Tarolli, P. A High-Resolution Physical Modelling Approach to Assess Runoff and Soil Erosion in Vineyards under Different Soil Managements. *Soil Tillage Res.* **2022**, *222*, 105418. [CrossRef]
7. Yang, S.; Zheng, X.; Zhao, G. Study on the Spatial Ecological Environment Effect and Influencing Factors of Ecological-Production-Living Spaces in the Guanzhong Plain Urban Agglomeration. *Arid Land Resour. Environ.* **2023**, *37*, 26–35. [CrossRef]
8. Liu, Q.; Lu, B.; Lin, W.; Li, J.; Lu, Y.; Duan, Y. An Improved Quantitative Analysis Method for the Unequal Supply and Demand of Ecosystem Services and Hierarchical Governance Suggestions. *Land* **2025**, *14*, 528. [CrossRef]
9. Zang, Z.; Zou, X.; Wu, L.; Song, Q.; Wu, X.; Yu, W. Evaluation of Ecological Well-Being and Ecological-Economic Efficiency in Mainland China Based on Equity and Efficiency Perspectives. *Acta Ecol. Sin.* **2017**, *37*, 2403–2414.
10. Xie, G.-d.; Chun-xia, L.; Yun-fa, L.; Du, Z.; Shuang-cheng, L. Ecosystem Assets Valuation of the Tibetan Plateau. *J. Nat. Resour.* **2003**, *18*, 189–196.
11. Costanza, R.; De Groot, R.; Sutton, P.; Van der Ploeg, S.; Anderson, S.J.; Kubiszewski, I.; Farber, S.; Turner, R.K. Changes in the Global Value of Ecosystem Services. *Glob. Environ. Change* **2014**, *26*, 152–158.
12. Deng, H.; Zhou, X.; Liao, Z. Ecological Redline Delineation Based on the Supply and Demand of Ecosystem Services. *Land Use Policy* **2024**, *140*, 107109.
13. Pascual, U.; Palomo, I.; Adams, W.M.; Chan, K.M.; Daw, T.M.; Garmendia, E.; Gómez-Baggethun, E.; De Groot, R.S.; Mace, G.M.; Martín-López, B. Off-Stage Ecosystem Service Burdens: A Blind Spot for Global Sustainability. *Environ. Res. Lett.* **2017**, *12*, 075001.
14. Rees, W.E. Ecological Footprints and Appropriated Carrying Capacity: What Urban Economics Leaves Out. In *The Earthscan Reader in Rural–Urban Linkages*; Routledge: London, UK, 2018; pp. 285–297.
15. Acuña, C.; Rodríguez, R.; Acuña, E. Revitalizing Transformed Landscapes: Economic Valuation of Artificial Lake Shorelines for Hydropower Generation. *Environ. Dev.* **2025**, *54*, 101163. [CrossRef]
16. Dam, M.M.; Durmaz, A.; Bekun, F.V.; Tiwari, A.K. The Role of Green Growth and Institutional Quality on Environmental Sustainability: A Comparison of Co2 Emissions, Ecological Footprint and Inverted Load Capacity Factor for Oecd Countries. *J. Environ. Manag.* **2024**, *365*, 121551.
17. Cheng, B.; Li, H.E.; Cheng, W.J.; Li, J.K. Fund Allocation Modeling of Compensation Subjects for Various Protection Levels of the Ecological Flow in Rivers under Runoff Variations. *J. Environ. Manag.* **2024**, *370*, 122690. [CrossRef]
18. Chen, Y.H.; Wu, Q.L.; Guo, L. Ecological Compensation Based on the Ecosystem Service Value: A Case Study of the Xin'an River Basin in China. *Water* **2024**, *16*, 2923. [CrossRef]
19. Xu, H.; Chen, L.X.; Li, Q.F. Research on Two-Way Ecological Compensation Strategy for Transboundary Watershed Based on Differential Game. *J. Environ. Manag.* **2024**, *371*, 123314. [CrossRef]
20. Guo, H.; Cai, Y.; Li, B.; Wan, H.; Yang, Z. An Improved Approach for Evaluating Landscape Ecological Risks and Exploring Its Coupling Coordination with Ecosystem Services. *J. Environ. Manag.* **2023**, *348*, 119277.
21. Shi, G. Ideas for the Construction of Ecological Compensation Mechanism in the Yellow River Basin in the Context of High-Quality Development. *Econ. Manag. Rev.* **2023**, *39*, 49–58. [CrossRef]
22. Chi, C.; Shen, J.Q.; Gao, X.; Hu, P.; Yi, P. Horizontal Ecological Compensation Standards Based on Ecosystem Services Flow. *Ecol. Indic.* **2024**, *163*, 112081. [CrossRef]
23. Niu, J.J.; Mao, C.M.; Xiang, J. Based on Ecological Footprint and Ecosystem Service Value, Research on Ecological Compensation in Anhui Province, China. *Ecol. Indic.* **2024**, *158*, 111341. [CrossRef]
24. Rega, C. Ecological Compensation in Spatial Planning in Italy. *Impact Assess. Proj. Apprais.* **2013**, *31*, 45–51. [CrossRef]
25. Fu, Y.Y.; Xiong, K.N.; Zhang, Z.Z. Ecosystem Services and Ecological Compensation of World Heritage: A Literature Review. *J. Nat. Conserv.* **2021**, *60*, 125968. [CrossRef]
26. Yang, S.; Zheng, X.Z. Spatio-Temporal Relationship between Carbon Emission and Ecosystem Service Value under Land Use Change: A Case Study of the Guanzhong Plain Urban Agglomeration, China. *Front. Environ. Sci.* **2023**, *11*, 1241781. [CrossRef]
27. Xie, G.; Zhang, C.; Zhang, L.; Chen, W.; Li, S. Improvement of Ecosystem Service Valorization Method Based on Unit Area Value Equivalent Factor. *J. Nat. Resour.* **2015**, *30*, 1243–1254.
28. Wang, Y.; Sun, X. Spatial and Temporal Evolution of Ecosystem Service Value and Influencing Factors in the Yellow River Basin. *Environ. Sci.* **2024**, *45*, 2767–2779. [CrossRef]
29. Niccolucci, V.; Bastianoni, S.; Tiezzi, E.; Wackernagel, M.; Marchettini, N. How Deep Is the Footprint? A 3d Representation. *Ecol. Model.* **2009**, *220*, 2819–2823.
30. Xiang, J.; Niu, J. Ecological Compensation and Comprehensive Zoning of Cultivated Land Based on Ecosystem Service Value and Extended Three-Dimensional Ecological Footprint Model: A Case Study of Shandong Province. *Land* **2025**, *14*, 316. [CrossRef]
31. Li, L.; Li, X.; Fan, H.; Lu, J.; Wang, X.; Zhai, T. Quantifying and Zoning Ecological Compensation for Cultivated Land in Intensive Agricultural Areas: A Case Study in Henan Province, China. *Land* **2024**, *13*, 1720. [CrossRef]
32. Ma, Y.H.; Chen, H.; Yang, M.M.; Aihemaiti, G.; Lu, W.J.; Zhao, R.J. Ecological Compensation Based on Multiscale Ecosystem Carbon Sequestration Service Flow. *J. Environ. Manag.* **2024**, *372*, 123396. [CrossRef]

33. Siche, R.; Pereira, L.; Agostinho, F.; Ortega, E. Convergence of Ecological Footprint and Emergy Analysis as a Sustainability Indicator of Countries: Peru as Case Study. *Commun. Nonlinear Sci. Numer. Simul.* **2010**, *15*, 3182–3192.
34. Wackernagel, M.; Rees, W. *Our Ecological Footprint: Reducing Human Impact on the Earth*; New Society Publishers: Gabriola Island, BC, Canada, 1998; Volume 9.
35. Report of the World Commission on Environment and Development: Our Common Future. 1987. Available online: http://www.un-documents.net/wced-ocf.htm (accessed on 15 January 2023).
36. Rees, W.E. Revisiting Carrying Capacity: Area-Based Indicators of Sustainability. *Popul. Environ.* **1996**, *17*, 195–215.
37. Bryan, B.A.; Gao, L.; Ye, Y.Q.; Sun, X.F.; Connor, J.D.; Crossman, N.D.; Stafford-Smith, M.; Wu, J.G.; He, C.Y.; Yu, D.Y.; et al. China's Response to a National Land-System Sustainability Emergency. *Nature* **2018**, *559*, 193–204. [CrossRef]
38. Zhang, X.R.; Wang, Y.S.; Yuan, X.F.; Shao, Y.J.; Bai, Y. Identifying Ecosystem Service Supply-Demand Imbalance for Sustainable Land Management in China?S Loess Plateau. *Land Use Policy* **2022**, *123*, 106423. [CrossRef]
39. Yang, S.; Su, H. Multi-Scenario Simulation of Ecosystem Service Values in the Guanzhong Plain Urban Agglomeration, China. *Sustainability* **2022**, *14*, 8812. [CrossRef]
40. Jia, Z.X.; Wang, X.F.; Feng, X.M.; Ma, J.H.; Wang, X.X.; Zhang, X.R.; Zhou, J.T.; Sun, Z.C.; Yao, W.J.; Tu, Y. Exploring the Spatial Heterogeneity of Ecosystem Services and Influencing Factors on the Qinghai Tibet Plateau. *Ecol. Indic.* **2023**, *154*, 110521. [CrossRef]
41. Bryan, B.A.; Ye, Y.Q.; Zhang, J.E.; Connor, J.D. Land-Use Change Impacts on Ecosystem Services Value: Incorporating the Scarcity Effects of Supply and Demand Dynamics. *Ecosyst. Serv.* **2018**, *32*, 144–157. [CrossRef]
42. Ouyang, Z.; Zheng, H.; Xiao, Y.; Polasky, S.; Liu, J.; Xu, W.; Wang, Q.; Zhang, L.; Xiao, Y.; Rao, E.M.; et al. Improvements in Ecosystem Services from Investments in Natural Capital. *Science* **2016**, *352*, 1455–1459. [CrossRef]
43. Zhang, L.F.; Fang, C.L.; Zhu, C.; Gao, Q. Ecosystem Service Trade-Offs and Identification of Eco-Optimal Regions in Urban Agglomerations in Arid Regions of China. *J. Clean. Prod.* **2022**, *373*, 133823. [CrossRef]
44. Hansen, R.; Frantzeskaki, N.; McPhearson, T.; Rall, E.; Kabisch, N.; Kaczorowska, A.; Kain, J.H.; Artmann, M.; Pauleit, S. The Uptake of the Ecosystem Services Concept in Planning Discourses of European and American Cities. *Ecosyst. Serv.* **2015**, *12*, 228–246. [CrossRef]
45. Zhao, Y.H.; Zhang, L.; Jia, X.; Mu, Q.; Han, L.; Liu, Z.; Zhang, P.; Zhao, M. Pattern and Trend of Ecosystem Service Value in the Loess Plateau of Northern Shaanxi. *Land* **2023**, *12*, 607. [CrossRef]
46. Deng, Y.J.; Yan, X.H.; Hou, M.Y.; Yao, S.B. Reassessing the Ecological Effectiveness of Ecological Restoration Programs: Evidence from a Quasi-Natural Experiment in China. *Ecol. Eng.* **2025**, *212*, 107506. [CrossRef]
47. Goodell, J.W.; Gurdgiev, C.; Karim, S.; Palma, A. Carbon Emissions and Liquidity Management. *Int. Rev. Financ. Anal.* **2024**, *95*, 103367. [CrossRef]
48. Siyiti, M.; Yao, X. Natural Resource Assets Management and Urban Carbon Emission Efficiency: Evidence from Quasi-Natural Experiment in China. *Energy Econ.* **2024**, *140*, 107963. [CrossRef]
49. Guo, X.M.; Fang, C.L.; Mu, X.F.; Chen, D. Coupling and Coordination Analysis of Urbanization and Ecosystem Service Value in Beijing-Tianjin-Hebei Urban Agglomeration. *Ecol. Indic.* **2022**, *137*, 108782. [CrossRef]
50. Zhao, C.; Liu, Y.; Yan, Z. Effects of Land-Use Change on Carbon Emission and Its Driving Factors in Shaanxi Province from 2000 to 2020. *Environ. Sci. Pollut. Res.* **2023**, *30*, 68313–68326. [CrossRef]
51. Ding, T.H.; Chen, J.F.; Fang, Z.; Wang, Y. Exploring the Differences of Ecosystem Service Values in Different Functional Areas of Metropolitan Areas. *Sustain. Prod. Consum.* **2023**, *38*, 341–355. [CrossRef]
52. Huang, C.B.; Zhao, D.Y.; Liu, C.; Liao, Q.P. Integrating Territorial Pattern and Socioeconomic Development into Ecosystem Service Value Assessment. *Environ. Impact Assess. Rev.* **2023**, *100*, 107088. [CrossRef]
53. Schmidt, S.; Hamidov, A.; Kasymov, U. Analysing Groundwater Governance in Uzbekistan through the Lenses of Social-Ecological Systems and Informational Governance. *Int. J. Commons* **2024**, *18*, 203–217. [CrossRef]
54. Wiedmann, T.; Lenzen, M.; Keysser, L.T.; Steinberger, J.K. Scientists' Warning on Affluence. *Nat. Commun.* **2020**, *11*, 31–107. [CrossRef]
55. Wan, R.X.; Qian, S.Y.; Ruan, J.H.; Zhang, L.; Zhang, Z.; Zhu, S.Y.; Jia, M.; Cai, B.F.; Li, L.; Wu, J.; et al. Modelling Monthly-Gridded Carbon Emissions Based on Nighttime Light Data. *J. Environ. Manag.* **2024**, *354*, 120391. [CrossRef] [PubMed]
56. Chen, Y.Z.; Zhang, S.S.; Yang, L.Z.; Zhang, X.C.; Yu, K.R.; Li, J. Ecological Footprint in Beijing-Tianjin-Hebei Urban Agglomeration: Evolution Characteristics, Driving Mechanism, and Compensation Standard. *Environ. Impact Assess. Rev.* **2024**, *109*, 107649. [CrossRef]
57. Liu, Z.; Guan, D.B.; Wei, W.; Davis, S.J.; Ciais, P.; Bai, J.; Peng, S.S.; Zhang, Q.; Hubacek, K.; Marland, G.; et al. Reduced Carbon Emission Estimates from Fossil Fuel Combustion and Cement Production in China. *Nature* **2015**, *524*, 335–338. [CrossRef] [PubMed]
58. Salzman, J.; Bennett, G.; Carroll, N.; Goldstein, A.; Jenkins, M. The Global Status and Trends of Payments for Ecosystem Services. *Nat. Sustain.* **2018**, *1*, 136–144. [CrossRef]

59. Wang, F.; Gu, N.; Dang, Y.T. Measuring Ecological Effects and Policy Mechanisms of Urban Land Use Transition—A Case Study of Xi'an, China. *Int. J. Urban Sci.* **2024**. [CrossRef]
60. Wang, Z.X.; Li, F.H.; Xie, Z.H.; Li, Q.Y.; Zhang, Y.L.; Dai, M.L. Decoupling CO_2 Emissions from Economic Growth in China's Cities from 2000 to 2020: A Case Study of the Pearl River Delta Agglomeration. *Land* **2023**, *12*, 1804. [CrossRef]

land

MDPI

Article

Empirical Analysis on the Mechanism of Industrial Park Driving Urban Expansion: A Case Study of Xining City

Ming Tian *, Zhuo He, Jinpeng Wei and Yicong Tian

School of Government, Beijing Normal University, Beijing 100875, China; hezhuozhuo@126.com (Z.H.); 202231240008@mail.bnu.edu.cn (J.W.); 202131240006@mail.bnu.edu.cn (Y.T.)

* Correspondence: tianm@bnu.edu.cn

Abstract: Taking Xining City as an example, this article analyzes the mechanism by which industrial park construction drives the expansion of urban population size and built-up area, based on a review of the process of industrial park development and urban population growth. It also discusses future urban governance models in light of urban development trends. The research finds: (1) In the process of urban development, industrial park construction is often the initial factor in the cumulative and cyclical development of a city; (2) As the level of development improves and the mode of economic growth changes, the government should timely adjust its strategies, shifting from the expansion of industrial park construction towards structural optimization and quality improvement. The most significant difference from previous research is that this paper emphasizes the importance of government planning. This study can not only demonstrate the general process of industrial parks promoting urban expansion, but more importantly, it explains the fundamental reasons for the transition of urban expansion to adjustment from a mechanism perspective, thereby eliminating the drawbacks of simply predicting urban scale evolution through data models.

Keywords: industrial park; urban scale; employment elasticity coefficient; urban governance

1. Introduction

Urban population serves as a crucial carrier of urban nature and function. Understanding and grasping the mechanisms and processes underlying changes in urban population size are foundational for urban managers to determine reasonable urban population sizes in urban planning, accurately judge and predict future urban trends, and more importantly, establish a rational urban governance way. In the research on population size prediction, major methods include the Logistic model [1], the Grey model [2], and the neural network model [3], as well as the Grey-neural network model [4–6] derived from the combination of the Grey model and the neural network model. These models all statistically analyze historical data on factors affecting changes in urban population size, explore their temporal variation patterns, and derive long-term trends in population size changes to predict their future variations [7]. These methods are mostly applicable to regions or entire countries with low cross-border migration speeds and small migration scales. For cities undergoing rapid urbanization, influenced by numerous factors, and experiencing significant uncertainty in population changes, model predictions based on historical data have been criticized for their accuracy [8–10]. More importantly, while models can output future evolution results, they cannot reveal the mechanisms and reasons behind these results. In the process of urban development and growth, apart from natural population changes, economic growth drives employment expansion, which attracts a large influx of migrant populations, further propelling population size growth. This increased population size, in turn, fuels urban economic growth, forming a cumulative cyclical process that is the fundamental mechanism sustaining urban population development.

According to the theory of cumulative cyclic causality, in a dynamic social process, there exists a cumulative cyclic causal relationship between various socioeconomic factors.

A change in one socioeconomic factor will induce a change in another socioeconomic factor, which, in turn, reinforces the initial change and leads the socioeconomic process to develop along the direction of the initial change, thereby forming a cumulative cyclical development trend [11]. Economic growth drives employment expansion, which facilitates labor transfer and agglomeration, further leading to population agglomeration in cities [12,13], promoting the expansion of urban population size and land use scale. Conversely, the growth of urban population size also drives economic growth. Since the reform and opening-up policy, numerous cities in China have established various types of development zones, industrial parks, and new towns, which have played a significant role in economic development. Many of these have served as "initial factors" at certain stages, driving rapid urbanization and becoming urban growth poles [14]. The role of various development zones, industrial parks, and even new urban areas in urban development is not difficult to understand, but there is a lack of systematic analysis and empirical research on how these new areas propel urban development. Secondly, since the 18th National Congress of the Communist Party of China, innovative and green development have become the themes of the era. Cities at all levels are gradually abandoning their original extensive development models and transforming and adjusting various industrial parks and economic development zones. After high-speed expansion and subsequent adjustment in industrial parks, it is necessary to delve into the processes and situations that urban development will undergo. Lastly, in the process of urban development, guided by different stages and goals, the concepts and methods of urban governance will change accordingly [15]. How should urban governance adjust to changes in the urban development process?

Taking Xining City, Qinghai Province, as an example, this paper analyzes the mechanism by which the construction of industrial parks drives urban development and urban population expansion, aiming to conduct a comprehensive and deep study on this issue. This study can not only demonstrate the general process of industrial parks promoting urban expansion, but more importantly, it explains the fundamental reasons for the transition of urban expansion to adjustment from a mechanism perspective, thereby eliminating the drawbacks of simply predicting urban scale evolution through data models.

2. Research Methods and Data of the Case Study Area

This study takes Xining City as an example to analyze the mechanism of how industrial park construction promotes urban population and land expansion based on the cumulative causation theory. Also known as the circular cumulative causation theory, this theory was first proposed by renowned economist Gunnar Myrdal in 1957 and later developed and formalized into models by Nicholas Kaldor, R.W. Dixon, and Gunnar Myrdal himself, among others. Myrdal and his colleagues believed that in a dynamic social process, there exists a circular and cumulative causal relationship among various socio-economic factors. A change in one socio-economic factor triggers a change in another, which in turn reinforces the initial change, leading the socio-economic process to evolve in the direction of the initial change, thus forming a cumulative and circular development trend [16–18]. Based on this concept, the analytical framework proposed in this paper is illustrated in Figure 1.

Another explanation for the driving force behind the development of industrial zones lies in the industrial clustering effect [19]. Clusters are formed by interconnected enterprises and institutions that are geographically concentrated and belong to similar industries within a specific field [20]. Industrial clusters are a common phenomenon in the process of industrialization, and various industrial clusters can be clearly seen in all developed economies [21]. The rise of industrial clusters stems from the discovery of positive externalities of agglomeration by scholars such as Marshall [22]. The agglomeration of enterprises in industrial clusters can build a stable regional innovation network [23], generating network effects and knowledge spillover effects, thereby driving the development of industrial parks. The coupling of internal and external networks within industrial agglomerations can bring about complementary and win-win results. The local innovation network is an essential guarantee for the spillover of relevant knowledge within the cluster, while the external global innovation network can, to a certain extent, bring more heterogeneous knowledge to the local cluster, thereby promoting the development of local innovation clusters [24]. Due to the similarity in cognition, technology, and other aspects brought about by geographical proximity, the internal connections within the cluster arise from the flow of homogeneous innovation knowledge with a high degree of relevance [25]. Related diversification is an important foundation for effective knowledge spillover [26].

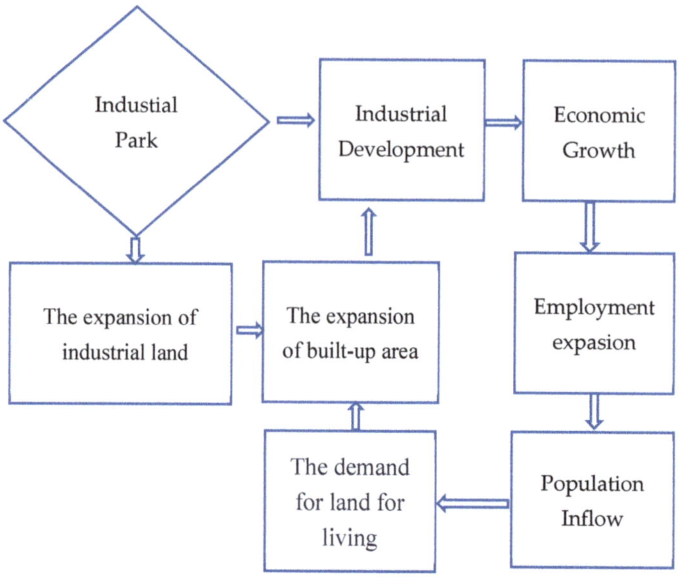

Figure 1. The analysis framework of industrial parks driving urban expansion.

Xining City, located on the eastern edge of the Qinghai–Tibet Plateau, in the eastern part of Qinghai Province, and within the river valley basin in the middle reaches of the Huangshui River, is the capital of Qinghai Province and one of the key cities in the Lanzhou-Xining urban agglomeration. It is an important city supporting the ecological security barrier of the Qinghai–Tibet Plateau and maintaining prosperity and stability in the northwest region. Xining City administers five districts and two counties, with a total municipal area of 7660 km^2. In 2020, the resident population was 2.468 million, with an urban population of 1.9406 million and an urbanization rate of 78.63%. The urban population in the central city area was 1.739 million, and the built-up area of the central city was 208 km^2. Xining is the only provincial capital city on the Qinghai–Tibet Plateau with an urban population exceeding one million and a built-up area exceeding 100 km^2. In 2020, the total GDP of Xining City was CNY 137.298 billion, accounting for 46.26% of the total GDP of Qinghai Province, with a per capita GDP of CNY 55,631, which is 13.58% higher than the provincial average.

Since 2000, within the central city area composed of five municipal districts including Chengzhong District, Chengdong District, Chengxi District, Chengbei District, and Huangzhong County (District), Xining City has successively established the Xining (National) Economic and Technological Development Zone, High-tech Industrial Development Zone (Biotechnology Industrial Park, located in Chengbei District), Chengnan New Area, Haihu New Area, Beichuan Industrial Park, and other new urban functional areas. Among them, the Xining (National) Economic and Technological Development Zone adopts a diversified management model within one zone, including the Dongchuan Industrial Park located in Chengdong District, the Ganhe Industrial Park located in Huangzhong County, and the Nanchuan Industrial Park located in Chengzhong District (see Figure 2). Driven by the development and construction of industrial parks and new urban areas, Xining City has entered a high-speed phase of economic and population growth as well as built-up area expansion, providing an excellent sample for observing and studying the expansion of population size and urban environmental changes driven by industrial parks.

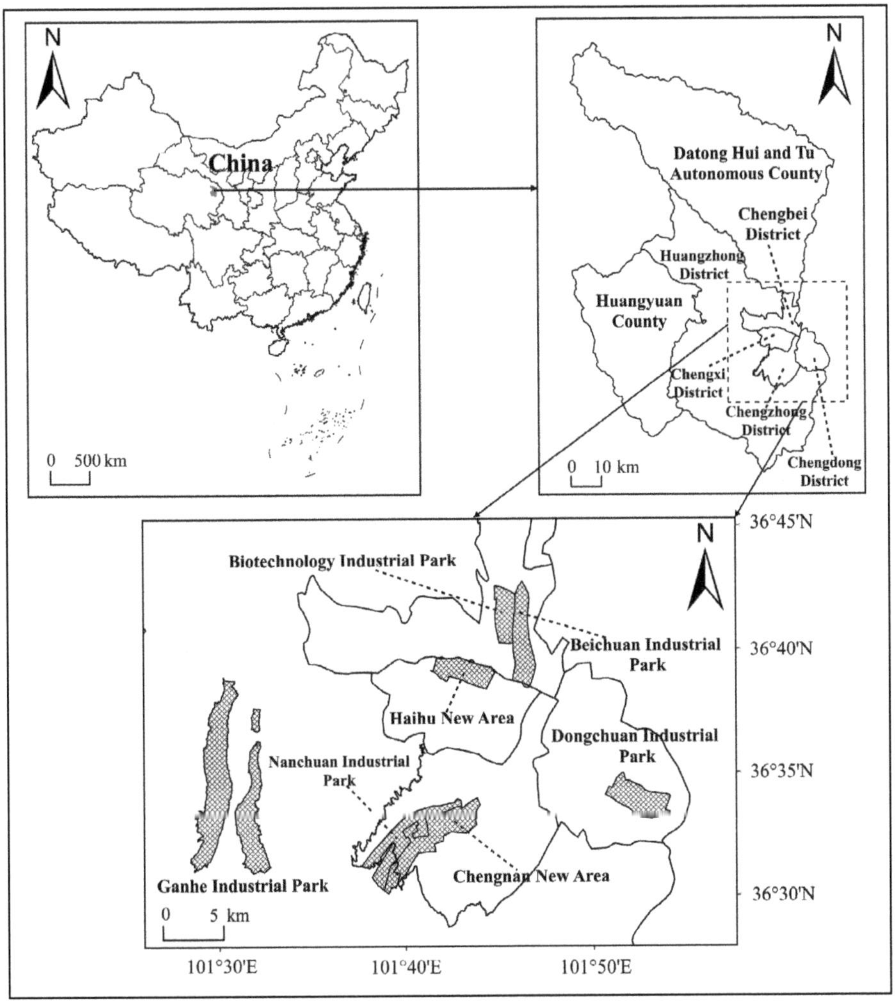

Figure 2. The layout of the central city area and industrial parks in Xining.

The population data for the main years in this article are primarily derived from the five population censuses conducted from the beginning of reform and opening-up until 2020. The population data, labor and employment data, and economic data for other years are sourced from the statistical yearbooks of Xining City from 1990 to 2021. In cases of inconsistent data within the statistical yearbooks, the data from the most recent year's statistical yearbook are used. Relevant information on industrial parks, new urban areas, and built-up areas is obtained from the Xining Development and Reform Commission, the Economic and Technological Development Zone Management Committee, as well as policy documents such as the "Twelfth Five-Year Plan for National Economic and Social Development of Xining City", the "Thirteenth Five-Year Plan for National Economic and

Social Development of Xining City", and the "Fourteenth Five-Year Plan for National Economic and Social Development of Xining City". All maps were produced by the authors using ArcGIS 10.5.

3. Industrial Park Construction and Urban Scale Expansion

3.1. Process of Industrial Park Construction and Urban Population Expansion

As a western inland province with ethnic minorities as the main population, compared to eastern regions and other parts of the country, Xining City implemented reform initiatives and related policies relatively late in the process of reform. In 1979, the Shekou Industrial Zone in Shenzhen was established, and in 1984, China's first batch of economic and technological development zones was newly established. By 1999, the total number of various urban new areas reached 815 [14]. However, as the capital city of Qinghai Province, Xining City was not approved to start construction of its first development zone, the Xining Economic and Technological Development Zone, until 2000. The development zone is located in Chengdong District, the eastern part of Xining City, Qinghai Province, and was renamed the Dongchuan Industrial Park of the Xining Economic and Technological Development Zone in 2006. Soon after, in April 2002, the Biotechnology Industrial Park was established in Chengbei District of Xining City, and in July 2002, the Ganhe Industrial Park was established in Huangzhong County (now Huangzhong District). The three industrial parks were unified and incorporated into the management system of the Xining Economic and Technological Development Zone in 2005 and 2006, forming a management framework of one zone with three parks. In February 2008, the Nanchuan Industrial Park of the Xining Economic and Technological Development Zone was established in Chengzhong District, forming a layout of one zone with four parks. In November 2010, the Biotechnology Industrial Park was upgraded to a national high-tech industrial development zone upon approval by the State Council and named the Qinghai High-tech Industrial Development Zone, becoming the only national high-tech zone in Qinghai Province. It was separated from the management system of the economic and technological development zone in 2017 for independent management.

As a carrier and engine driving rapid economic development, industrial parks have gradually played their role. In the 10 years before 2000, Xining City's average annual GDP growth rate was 7.56%, with the highest year being 1998 at 10.1%. In other years, it did not exceed double digits and was not significantly different from the economic growth rate of Qinghai Province (see Figure 3). However, Xining's economic growth continued to rise and gradually surpassed the provincial economic growth rate, reaching a peak of 18.2% in 2010. It maintained a double-digit growth rate for 16 consecutive years until it dropped to 9.8% in 2016, far exceeding the economic growth rate of Qinghai Province. This period roughly coincides with the development and construction of industrial parks, and the contribution of industrial parks to economic growth has gradually increased. In 2007, the industrial added value of the four major industrial parks accounted for 25.7% of the city's total industrial added value. By 2009, the proportion exceeded 50% for the first time, reaching 50.51%, and peaked at 73.9% in 2016. After 2016, although the proportion of industrial added value from the parks in the city began to decline, it still accounted for about 55% by 2019, before the outbreak of the pandemic (see Figure 4).

Rapid economic growth inevitably generates significant demand for labor. During the "12th Five-Year Plan" period from 2006 to 2010, Xining City created 95,000 new urban jobs. During the subsequent "12th Five-Year Plan" period from 2011 to 2015, the cumulative number of new urban jobs reached 150,400. During the "13th Five-Year Plan" period from 2016 to 2020, the cumulative number of new urban jobs was 159,000, and by 2020, the number of urban employees in Xining City surpassed one million, reaching a high of 1.04 million. Correspondingly, the urban population increased from 910,000 in 2000 to 1.1483 million in 2010 and reached 1.688 million in 2020, with an increase of 778,000 urban residents over the past 20 years. In contrast, during the entire 90s of the last century, the urban population of Xining City increased by only 260,000 (see Figure 5).

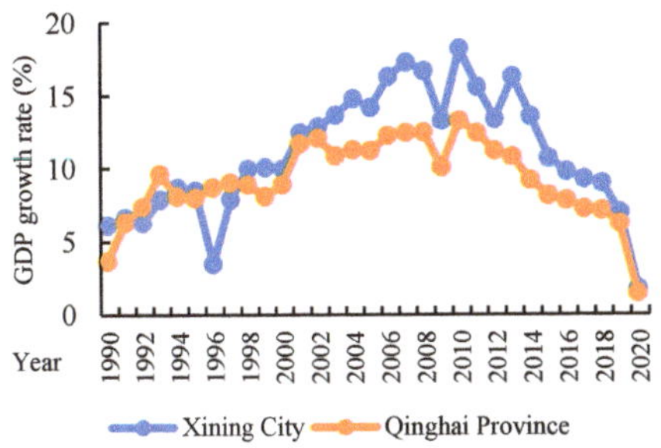

Figure 3. The comparison of GDP growth rate between Xining City and Qinghai Province.

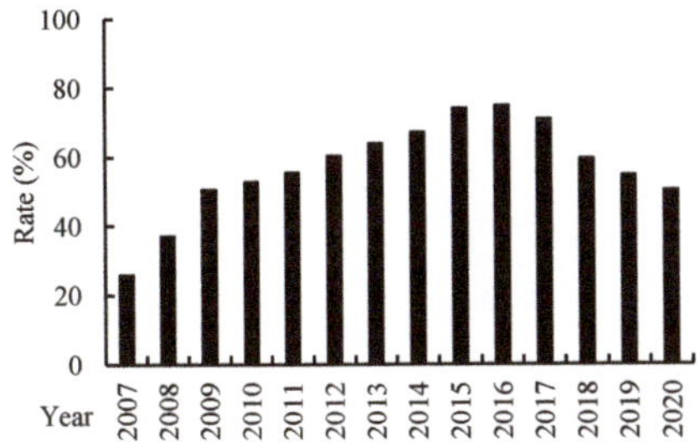

Figure 4. The value added of industry in industrial parks as a share of the city's value added of industry.

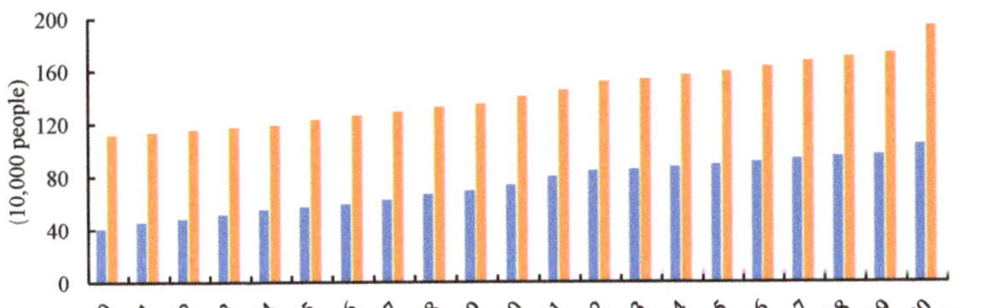

Figure 5. Changes in urban employees and urban population in Xining City from 2000 to 2020.

3.2. *Process of Urban Built-Up Area Expansion*

The construction of industrial parks and new cities relying on these parks has driven rapid expansion of Xining's urban area. The built-up area expanded from 52 km^2 in the early 2000s to 91 km^2 in 2010 and reached 208 km^2 by 2020 (see Figures 6 and 7), representing an increase of 108% over the past decade. In contrast, during the entire 90s of the last century, the built-up area of Xining increased by only 2 km^2. The expansion of the built-up area has been achieved through two modes, where one is the expansion driven by the construction of industrial parks themselves. The Dongchuan Industrial Park, the first established in Xining, has a planned area of 12.79 km^2; the Biotechnology Industrial Park initially had a planned construction area of 10 km^2; the Ganhe Industrial Park has a planned area of 10.89 km^2; and the Nanchuan Industrial Park has a planned area of approximately 10 km^2. By around 2010, except for the Dongchuan Industrial Park due to terrain constraints, the other three industrial parks had all undergone expansion, with the Biotechnology Industrial Park expanding by 23.5 km^2, Nanchuan expanding to 31.39 km^2, and Ganhe Industrial Park reaching 35.28 km^2. The planned area of the parks increased from the initial 53.78 km^2 to 101.06 km^2.

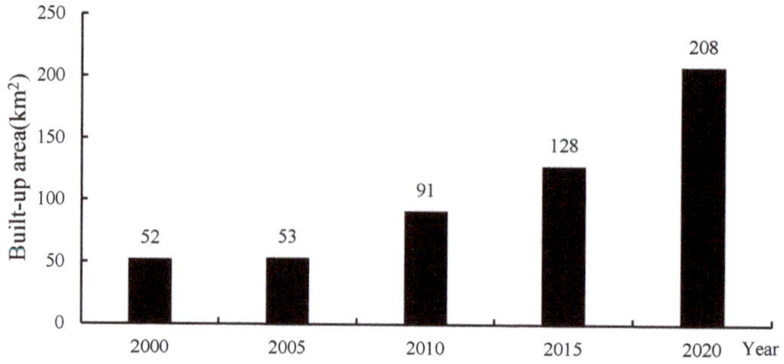

Figure 6. The expansion process of the built-up area of Xining City center.

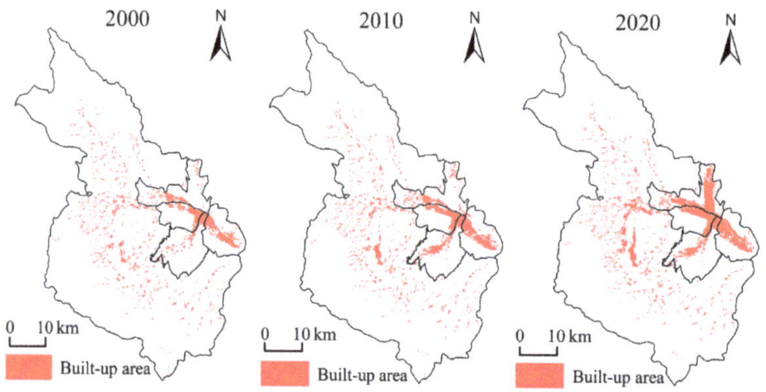

Figure 7. The process of spatial expansion of construction land in Xining City.

With the development of information technology, the dominant type of industrial parks has gradually shifted from suburban or suburban-like park models to urban models with more "urban characteristics" [15]. As the urban population grows, new city construction has become another major pathway for the expansion of Xining's built-up area. When the Ganhe Industrial Park underwent its second expansion in 2008, it also initiated the Kangchuan New City relocation and resettlement project involving 26,000 people. As the largest relocation and resettlement project in Qinghai Province at that time, it was equipped with medical, educational, commercial, cultural, and other living service facilities. The construction of the Ganhe Industrial Park also facilitated the transformation of Huangzhong County into a district, becoming part of Xining's urban area in 2019—Huangzhong District. During the construction of the Nanchuan Industrial Park and the Ganhe Industrial Park, Xining developed the Chengnan New Area between the Nanchuan Industrial Park in the southern part of the city and the Ganhe Industrial Park, with a total area of 30 km^2. In 2007, construction of the Haihu New City began in the Chengxi District, covering an area of 10.5 km^2 and planned to accommodate a population of 150,000. In 2014, relying on the Biotechnology Industrial Park (Qinghai National High-tech Industrial Development Zone), the Beichuan New Area was officially developed and constructed, with a planned area of 8.6 km^2. Many new campuses of universities such as Qinghai University and Qinghai Normal University are located in the new area. The construction of three new cities relying on the three major industrial parks in the south, north, and west has basically completed the expansion of Xining's central urban built-up area, forming the current built-up area of 208 km^2.

From the perspective of the entire process of Xining's urban development, industrial parks have been the "initial factor" driving urban development since 2000. This "initial factor", under the cumulative and circular causal law, has brought rapid economic growth and increased the urban employment capacity, thereby driving the expansion of the urban population and land use. The expansion of the city scale, in turn, promotes economic growth, creating a virtuous cycle. This process of development and expansion is summarized and illustrated in the following diagram (see Figure 8).

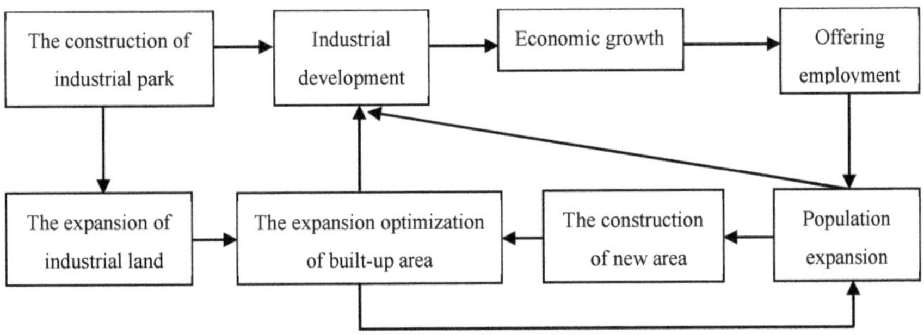

Figure 8. The cumulative cyclic process of urban scale expansion driven by the construction of industrial parks in Xining City.

4. Transformation of Economic Growth Patterns and Trends in Urban Evolution

4.1. Adjustment and Transformation of Economic Growth Patterns

The rapid expansion of urban population size and land use scale will inevitably put pressure on the ecological environment, leading to sustainable development issues [26]. Since the 18th National Congress of the Communist Party of China, the country has increasingly emphasized the quality of economic development rather than solely emphasizing the speed of development. Economic development has shifted from the unsustainable model that relied on extensive resource input to a sustainable development model mainly driven by technological progress and improvements in labor productivity. On the one hand, the economy has entered a new stage of medium-to-low-speed development from high-speed growth. Since 2013, Xining's economic development has entered a downward trend, necessitating the exploration of new economic growth points. On the other hand, the pressure on the ecological environment caused by large-scale development has prompted the city to adjust its development strategy. In the "12th Five-Year Plan", the Xining municipal government proposed to "increase industrial structural adjustments, significantly reduce energy and resource consumption and environmental pollution through technological innovation and technological transformation, gradually form a sustainable development model, establish and improve the characteristic industrial system, and embark on a path of new industrialization". During the "13th Five-Year Plan" period, Xining intensified the adjustment of industrial parks, proposing that each industrial park should focus on its leading industries, promote industrial upgrading, rationally divide labor in space, improve land use efficiency, and form agglomeration advantages: "Dongchuan focuses on silicon material photovoltaics and light alloy materials; Nanchuan focuses on lithium-ion battery new energy and Tibetan carpet and woolen textile industry; Chengbei focuses on biomedicine and equipment manufacturing; Ganhe focuses on non-ferrous metal deep processing and specialty chemicals; Beichuan focuses on high-tech industries". The recent "14th Five-Year Plan" identifies green development and innovation development as the two core pillars of Xining's development: "Accelerate the formation of a system, industrial system, and production and lifestyle that match ecological protection, follow the path of green development, and build an ecological civilization city; adhere to the core position of innovation in the overall situation of modernization, accelerate the construction of a modern industrial system, and actively promote high-quality development".

The shift from past extensive development to the pursuit of high-quality development will reshape the trajectory of urban development. Improvements in labor productivity will further reduce the employment elasticity coefficient of economic growth, lower the labor absorption capacity, and thereby reduce the speed of urban population agglomeration. At the same time, due to the decline in the birth rate and the increase in the aging population, although the child dependency ratio of the working-age population has decreased, the elderly dependency ratio has increased. The migrant population is the main source of population growth in central cities, and changes in the bringing-family coefficient will also be an important factor affecting future urban scale changes.

4.2. Changes in the Employment Elasticity Coefficient of Economic Growth

Different economic growth patterns and different industrial types have different labor demand intensities and labor intensity at different stages of development, leading to varying abilities to drive new employment through economic growth. Since 2013, Xining's economy has entered a downward trend, and along with the decline in development speed, the scale of new employment brought by GDP growth has also shown a downward trend, manifested as a decline in the overall employment elasticity coefficient. From 2000 to 2013, the employment elasticity coefficient of economic growth in Xining was above 0.3 in most years, with the highest reaching 0.55 in 2011. After 2012, the employment elasticity coefficient gradually declined, with no year exceeding 0.3, and the highest in 2016 was only 0.27 (see Figure 9). Especially for the secondary industry, with industrial parks as the main carrier, after experiencing scale expansion, the number of employees peaked in 2011 and 2012 and then tended to decline (see Figure 10), contributing most significantly to the decline in the employment elasticity coefficient. It can be seen here that the decline in the employment elasticity coefficient is not only driven by the government's proactive response and active promotion but also an objective requirement for industrial upgrading and development to a certain extent.

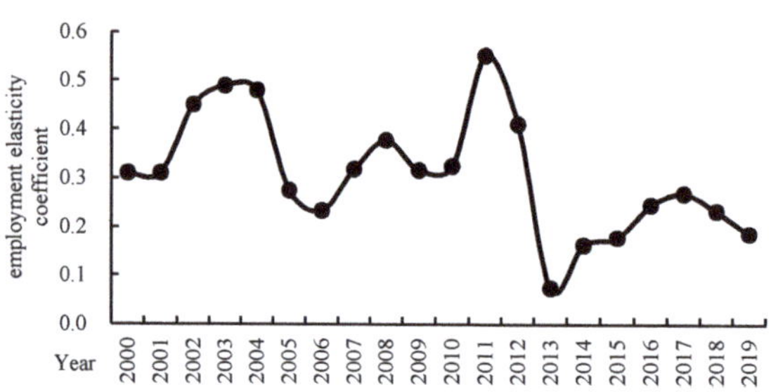

Figure 9. Changes in the employment elasticity coefficient of economic growth in Xining City.

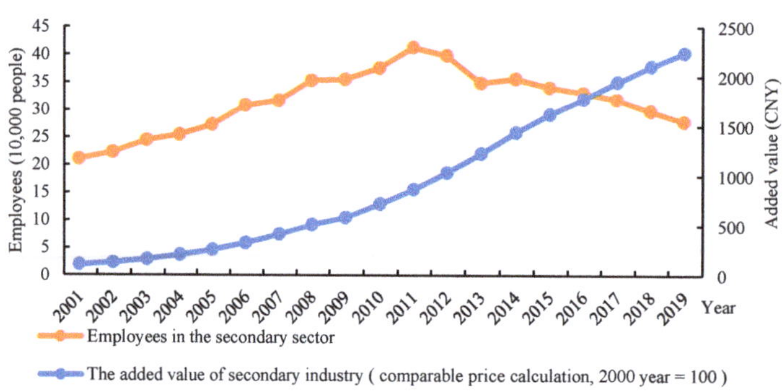

Figure 10. The added value of the secondary industry and the change of employees in Xining City.

4.3. Dependency Ratio and Changes in Urban Population Size

The urban economy is an economic system primarily based on the secondary and tertiary industries. Employment expansion in cities leads to spatial migration and agglomeration of population, thereby driving the expansion of urban size and being the primary mode for the growth and development of urban population size. In terms of net population migration, since 2000, the net migration of population in the central urban area of Xining has been continuously increasing, particularly evident in Chengdong District and Chengbei District. The net migration of population in the four central districts increased from 295,000 in 2010 to 522,700 in 2020 (see Table 1). The fundamental purpose of migration for non-household registered individuals is to secure employment and earn income. The results of net population migration in different districts and counties of Xining demonstrate the relationship between employment expansion and urban population growth. Although the employment elasticity coefficient of economic growth in Xining declined in the latter half of the past 20 years, the expansion of economic and employment scales still significantly contributed to population growth. Furthermore, the magnitude and trend of urban population growth, along with changes in the dependency ratio resulting from demographic structural evolution, are also important factors influencing employment-driven population changes.

Table 1. Net migration of population in Xining City counties and districts in major years (10,000 people).

Year	Chengdong District	Chengzhong District	Chengxi District	Chengbei District	Huangzhong District	Datong County	Huangyuan County
1990	2.27	—	−3.89	—	0.31	0.41	0.05
2000	9.11	—	4.91	4.65	0.21	1.51	0.02
2010	11.1	6.62	3.26	8.56	−2.71	−1.04	−0.07
2020	22.56	7.23	5.4	17.08	−8.64	−2.66	−1.85

The dependency ratio in Xining has experienced a trend of first decreasing and then increasing. It declined from nearly 2.5 in the early 21st century to around 2.0 in 2008 and further dropped to its lowest point in 2016. Since 2016, it has gradually risen over the past five years, forming a "U"-shaped trend (see Figure 11). This changing trend has solid practical implications and is influenced by changes in the population's age structure. As shown in Figure 10, the total dependency ratio in Xining was 47.85 in 2000 and then declined to 40.52 in 2010. The primary reason for this decline was the significant drop in the child dependency ratio due to the decrease in the birth rate. In 2020, the total dependency ratio rose again to 43.82, primarily due to the impact of population aging, where the elderly dependency ratio increased significantly while the child dependency ratio remained relatively stable (see Figure 12). The evolution of the population structure is the root cause of changes in the dependency ratio. The changes in Xining indicate that with the adjustment of China's family planning policy, while the birth rate and child dependency ratio remain stable, the increasing trend of aging will lead to a rise in the dependency ratio. This will be an important factor in judging the future changes in urban scale in China.

Another phenomenon emerging in the stage of high-quality development is the continuous optimization of urban land use structure, the improvement of public facilities, and the enhancement of urban space. In the process of urban population expansion, the expansion of built-up areas into urban districts has provided possibilities for incorporating larger green spaces and conditions for the renovation of old urban areas. In 2020, Xining established 124 parks and street green spaces. The green space ratio in built-up areas increased from 18.7% in 2000 to 35.6% in 2010 and 40.02% in 2020. The per capita public green space area rose from 3.94 m^2 to 12.82 m^2 (see Table 2). The improvement in the livability of the urban environment has increased its attractiveness, which is one of the factors driving population agglomeration. In the 14th Five-Year Plan, the Xining Municipal Government further proposed to "build a high-level park city".

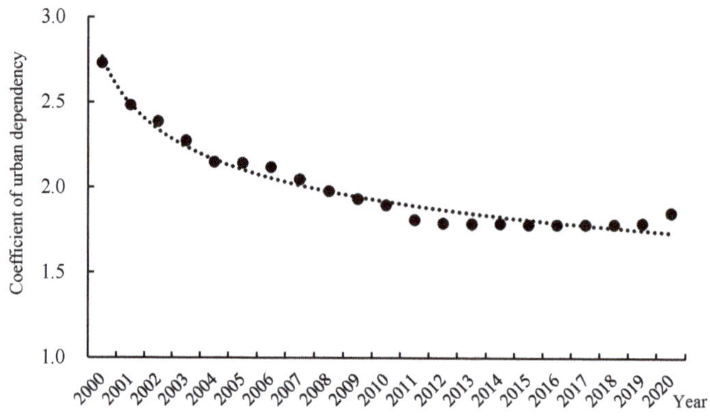

Figure 11. Coefficient of urban dependency.

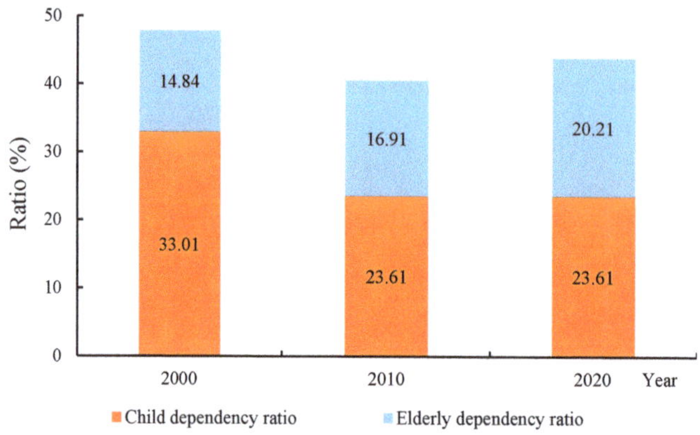

Figure 12. Changes in the population dependency ratio in Xining.

Table 2. Green space rate and public green space per capita in built-up areas of Xining City from 2000 to 2020.

Year	2000	2005	2010	2015	2020
Road area per capita (m^2)	6.48	6.81	7.35	7.8	12.9
Green space rate of Built-up areas (%)	18.7	28.0	35.6	39.02	40.02
Public green space per capita (m^2)	3.94	6.16	9.00	12.0	12.82

In the process of transforming economic growth from an extensive mode to a high-quality one, the decline in economic growth rate and the functional adjustment of industries and industrial parks have led to an increase in labor productivity, with a decreasing employment elasticity coefficient as a long-term trend. Along with the improvement of economic development, the decline in the birth rate and the deepening of aging are also long-term trends in social development, driving the dependency ratio to first decrease and then increase. The dual overlapping effects of the dependency ratio and the employment elasticity coefficient mean that compared to the period of extensive development, the growth rate of urban population will slow down, leading to a slower pace of land expansion and ushering in a period of internal optimization and adjustment for urban space. This process can be illustrated by Figure 13.

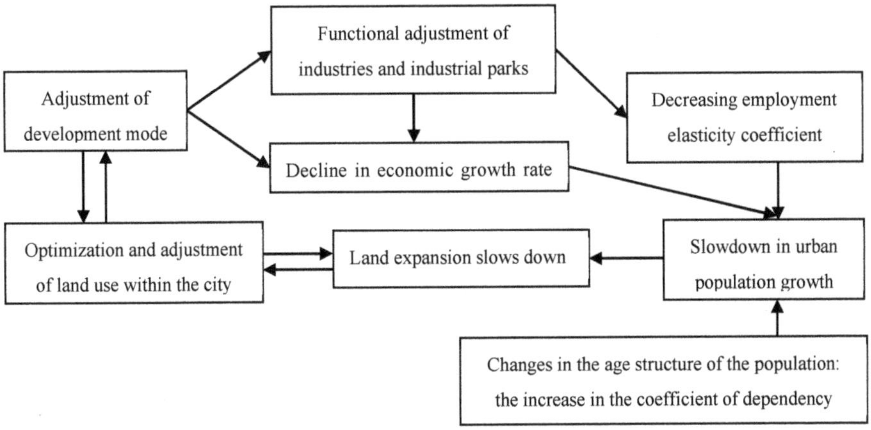

Figure 13. Relationship between economic growth and urban population growth at the stage of high-quality development.

5. Conclusions and Discussion

Industrial parks are crucial factors driving urban economic growth and urban expansion. This study finds that the construction of industrial parks is often the "initial factor" in the cumulative and cyclical development process of cities. Firstly, in the urban development process, the construction of industrial parks drives the economy to a new growth platform, expands the capacity to absorb labor, drives population agglomeration, and provides impetus for the construction of new urban areas. The construction of new urban areas and the expansion of industrial parks become the two major ways for the expansion of built-up areas, marking the first wave of urban development driven by industrial parks. Secondly, with the improvement of the development level, the transition from an extensive development mode solely pursuing economic growth rate to a higher-quality development mode promotes the adjustment of industrial parks, the reorientation of industrial development, and leads to a decline in economic growth rate and the employment elasticity coefficient of economic growth. On this basis, coupled with changes in the dependency ratio caused by demographic structural changes, it results in a decrease in urban population growth and urban construction land expansion. At this point, the city enters a stage of internal land optimization and adjustment, accompanied by the improvement of urban infrastructure and the gradual increase in urban per capita road area and per capita green area, marking the second wave of intensive urban development. The understanding of the process and mechanism of urban expansion driven by the construction, development, and adjustment of industrial parks, as well as the changes in the employment elasticity coefficient and the dependency ratio brought by demographic structural changes, is a key to predicting urban population size. The "initial factor" or the dominant factor leading to development and change may vary in different cities, but understanding the process and mechanism and recognizing the development patterns of the studied city should be the premise for future predictions.

After the second phase of optimization and adjustment, during the pandemic years of 2021, 2022, and 2023, Xining's GDP growth rates reached 8.1%, 2.1%, and 8.6%, respectively. Notably, the GDP growth rates in 2021 and (hypothetically) 2023 surpassed the provincial average by 2.4 and 3.4 percentage points, respectively. This underscores the role that Xining's optimization of industrial parks has played in driving economic recovery post-pandemic. In 2023, China's urbanization level reached 66.2%, entering the middle to late stages of urbanization. Based on the two-stage process of urban scale development driven by industrial parks, promoting urban intensification and sustainable development is not only a necessity for the situation and practical development but also supported by the theory of urban development patterns. In the future, with the continuation of the new normal of economic growth, labor productivity will increase with the improvement of the development level, the employment elasticity coefficient will decline, and the trend of large-scale urban expansion across the country will further reverse. Both urban population and land scale will enter a period of stable development. The focus of urban construction will shift from expanding urban scale to enhancing urban quality [27], and correspondingly, urban governance concepts and practices must also adapt to this transformation [15]. During the period of rapid urban expansion, the "incremental" construction of industrial parks and new towns must be addressed through an "incremental" governance model, expanding the spatial scope of urban governance and transitioning from a rural governance model to an urban governance model. When cities enter the "stock" construction period of stable development, the emphasis is on the optimization and upgrading of "stock", with "stock" governance as the main focus, targeting the construction of livable, innovative, smart, green, humanistic, and resilient cities [28], and improving the scientific and refined level of urban governance with people as the core.

While industrial parks cannot explain all the reasons for urban expansion, they are a significant force and one of the key paths for promoting urban development. As a city in the underdeveloped western region of China, the study of Xining shows that for economically underdeveloped areas, following the conventional path of development is challenging to achieve economic "catch-up" and reverse their lagging status. However, through the concentrated development of industrial parks, improving the local investment environment, and enhancing infrastructure quality, it becomes possible to achieve catch-up growth. Nevertheless, the construction of industrial parks is a gradual process that must be coordinated with national and even global macroeconomic trends, avoiding unlimited development that could lead to disorderly expansion or even the emergence of "ghost cities". Strategies must be flexibly adjusted according to different development stages and changes in macroeconomic conditions, with comprehensive support in land development and consolidation, infrastructure construction, labor and social security, and other aspects. All industrial parks in Xining are under the unified management of the Xining Municipal Government. To avoid vicious competition among different industrial parks, the government has conducted precise positioning and overall planning for the functions of each park. This highlights the critical importance of coordinating the positioning of different industrial parks within the same region.

The previous research on industrial districts and clusters emphasized the innovation networks formed within clusters by enterprises, as well as the knowledge sharing and spillover effects that arise from multiple similarities. These effects emerged spontaneously among enterprises. However, the most significant difference from previous studies lies in the emphasis placed on the importance of government planning in this paper. Specifically, it highlights that, especially in economically underdeveloped regions, governments should seize opportunities in a timely manner based on external environments and internal development needs, proactively promote the construction of industrial parks, and adjust the development strategies of industrial districts according to external and internal changes. This proactive approach is crucial for regional economic development.

Author Contributions: M.T. designed research and wrote the paper; Z.H. revised and proofread the entire paper; J.W. and Y.T. calculated the data and designed and produced figures and tables. All authors have read and agreed to the published version of the manuscript.

Funding: This study was supported by the Second Tibetan Plateau Scientific Expedition and Research Program (2019QZKK0406) and the National Natural Science Foundation of China (42371197).

Data Availability Statement: The data presented in this study are available upon request from the corresponding author. The data are not publicly available because the research team signed a data confidentiality agreement with the local government.

Conflicts of Interest: The authors declare no conflicts of interest.

References

1. Xie, T.; Zhu, M.; Xu, H. Research on the population structure of the family planning policy adjustment. *China Popul. Resour. Environ.* **2015**, *25*, 122–124.
2. Gao, H.; Yao, T.; Kang, X. Population forecast of Anhui province based on the GM (1, 1) model. *Grey Syst. Theory Appl.* **2017**, *7*, 19–30. [CrossRef]
3. Riiman, V.; Wilson, A.; Milewicz, R.; Pirkelbauer, P. Comparing artificial neural network and cohort-component models for population forecasts. *Popul. Rev.* **2019**, *58*, 2. [CrossRef]
4. Xu, L.; Li, H.; Li, J. Research on population prediction based on grey prediction and radial basis function network. *Comput. Sci.* **2019**, *46*, 431–435.
5. Li, G.; Wu, T.; Xu, S. Prediction model of population gross based on grey artificial neural network and its application. *Comput. Eng. Appl.* **2009**, *45*, 215–218.
6. Wang, Y.; Xie, Y.; Liu, R.; Zhang, H. Population prediction and resource allocation in megacities from the optimum population perspective: A case study of Beijing, Shanghai, Guangzhou and Shenzhen. *Acta Geogr. Sin.* **2021**, *76*, 352–366. [CrossRef]
7. Zhou, R. Application of the grey model to forecasting scale of urban population. *J. Inn. Mong. Norm. Univ.* **2005**, *34*, 81–83.
8. Chen, Y.; Liu, T. Reflection on population prediction in Beijing urban master plan. *Planners* **2015**, *31*, 16–21.
9. Li, Z. On the relationship between urban population prediction and resource and environment capacit. *J. Guangzhou Univ.* **2001**, *5*, 81–83+91.
10. Huang, R. Thoughts on population forecasting. *Popul. Res.* **2004**, *66*, 88–90.
11. Myrdal, G.; Sitohang, P. *Economic Theory and Under-Developed Regions*; Duckworth: London, UK, 1957; pp. 10–48.
12. Henderson, V.; Kuncoro, A.; Turner, M. Industrial development in cities. *J. Political Econ.* **1995**, *103*, 1067–1090. [CrossRef]
13. Xiang, B.; Yang, X. Research on the effects of the Chinese urban amenities on the growth of urban population. *Urban Dev. Stud.* **2017**, *24*, 146–150.
14. Zhou, C. *China's New Town*; Science Press: Beijing, China, 2021; pp. 1–87.
15. Cao, H.; Huo, W. The paradigm transformation about theory of urban governance and its enlightenment to China. *Chin. Public Adm.* **2013**, *7*, 94–99. [CrossRef]
16. Gunnar Myrdal. *An American Delamma*; Harper Brothers Publishing House: New York, NY, USA, 1944.
17. Veblen, T.B.; Mallock, W.H. Aristocracy and Evolution:A Study of the Rights, the Origins and the Social Functions of the Wealthier Classes. *J. Political Econ.* **1898**, *6*, 430–431. [CrossRef]
18. Argyrous, G. Setterfield on Cumulative Causation and Interrelatedness:A Comment. *Camb. J. Econ.* **2001**, *25*, 103–106. [CrossRef]
19. Tian, M.; Fan, J. New Industrial Districts'Developmental Mechanism and Comparative Analysis to Traditional Spatial Organization Theories. *Prog. Geogr.* **2003**, *22*, 186–194.
20. Poter, M. Clusters and the new economics of competition. *Harv. Bus. Rev.* **1998**, *76*, 77–90.
21. Poter, M. *The Competitive Advantage of Nations*; Free Press: New York, NY, USA, 1990.
22. Zhou, C.; Zeng, G. Progress and prospect of international research on industrial cluster: A perspective from Economic Geography. *Econ. Gepgraphy.* **2018**, *38*, 11–19.
23. Udo Staber. The Structure of Networks in Industrial Districts. *Int. J. Reg. Res.* **2001**, *25*, 538–552.
24. Bathelt, H.; Malmberg, A.; Maskell, P. Clusters and knowledge: Local buzz, global pipelines and the process of knowledge creation. *Prog. Hum. Geogr.* **2004**, *20*, 31–56. [CrossRef]
25. Tanner, A.N. The emergence of new technology-based industries: The case of fuel cells and its technological relatedness to regional knowledge bases. *J. Econ. Geogr.* **2015**, *16*, 611–635. [CrossRef]
26. Frenken, K.; Boschma, R.A. A theoretical framework for evolutionary economic geography: Industrial dynamics and urban growth as a branching process. *J. Econ. Geogr.* **2007**, *7*, 635–649. [CrossRef]
27. Wei, J.P.; Tian, M.; Wang, X. Spatiotemporal variation in land use and ecosystem services during the urbanization of Xining City. *Land* **2023**, *12*, 1118. [CrossRef]
28. The Central People's Government of the People's Republic of China. *The Outline of the 14th Five-Year Plan for Economic and Social Development and Long-Range Objectives through the Year 2035 of the People's Republic of China*; People's Publishing House: Beijing, China, 2021; pp. 10–57.

land

MDPI

Article

How Resource-Exhausted Cities Get Out of the Innovation Bottom? Evidence from China

Zihan Hu [1], Min Wu [2], Dan Yang [2], Tao Luo [3] and Yihao Tian [2,4,*]

[1] Business School, The University of New South Wales, Sydney, NSW 2052, Australia; zihan.hu@student.unsw.edu.au

[2] School of Public Administration, Sichuan University, Chengdu 610041, China; wuminhelen@163.com (M.W.); yang_d@stu.scu.edu.cn (D.Y.)

[3] Business School, Sichuan University, Chengdu 610064, China; taoluo@stu.scu.edu.cn

[4] Social Development and Social Risk Control Research Center of Sichuan Philosophy and Social Sciences Key Research Base, Chengdu 610065, China

* Correspondence: yihaotian@scu.edu.cn

Abstract: The transformation and upgrading of resource-exhausted cities are crucial for regional sustainable development, but how to help them overcome innovation challenges remains to be explored. Based on data from 2003 to 2016, this study used a difference-in-differences (DID) method to examine the impact of China's support policy for resource-exhausted cities on urban innovation and tests for long-term mechanisms. The results indicate that the support policy significantly enhanced regional innovation levels. The mechanism tests showed that these policies promoted urban innovation through long-term mechanisms of increasing marketization and upgrading industrial structures. Further analysis revealed that the innovation-promoting effects of the policies were more significant in resource-exhausted cities located in the eastern region, those not dependent on coal, those with a low reliance on extractive industries, and those with a favorable talent environment. The findings suggest that the government should provide policy support to achieve the transformation, upgrading, and sustainable development of resource-exhausted cities through urban innovation.

Keywords: resource-exhausted cities; sustainable urban development; urban innovation; marketization; China

1. Introduction

For a long time, developing countries have hindered sustainable development by trading high resource consumption for low economic growth [1]. In the early stages of national industrialization, resource-based cities relied on their abundant natural resources to transfer surplus value to other regions through low resource prices, contributing significantly to national development [2]. However, due to the constraints of limited resources and the laws of urban development, resource-based cities inevitably go through a process of construction, prosperity, decline, and transformation [3]. As resources are gradually depleted, the economic benefits of the resource industry diminish, and the development of these resource-based cities reaches its end stage, becoming known as resource-exhausted cities [4]. Thus, the transformation of resource-exhausted cities is a significant issue in the economic and social development of countries worldwide, including developed nations like Germany's Ruhr area and France's Lorraine region.

Contrary to the "convergence process" in Solow's growth theory, empirical evidence shows that abundant resources have not benefited regions; instead, they have constrained local economic development. This dilemma faced by resource-exhausted cities is termed the "resource curse" [5–7]. The "resource curse" refers to regions rich in resources becoming overly dependent on a single economic structure, resulting in low industrialization and difficult industrial transformation [8], as exemplified by Venezuela. International experience shows that financial aid can only address the surface issues of resource-exhausted cities. To transform from resource-dependent to innovation-driven economies and achieve sustainable development, it is crucial to resolve the deeper contradictions of lacking long-term economic development mechanisms [9].

Innovation is a key driver of economic transformation and development. According to endogenous growth theory, sustained economic growth cannot rely on external forces, and endogenous technological progress is the determinant of long-term economic growth [10]. Urban innovation has become a critical driver of sustainable urban economic development. Concentrating talents, enterprises, capital, and other innovation elements, urban innovation is a significant carrier of national innovation activities and refers to the processes and products centered on urban technological progress [11,12]. For countries worldwide, exploring how to enhance urban innovation capacity is of great importance in promoting national innovation systems and sustainable economic development. Ample theoretical and empirical evidence suggests that urban innovation plays a vital role in cultivating new economic growth points, optimizing and upgrading industrial structures, improving resource utilization efficiency, enhancing urban resilience, and promoting inclusive social development [13–16]. However, resource-exhausted cities often lag in innovation. Long-term dependence on single resource-based industries leads to low innovation levels, while poor economic development further reduces innovation factors [17]. Effectively improving the innovation level of resource-exhausted cities is an important issue that is worth exploring.

As the world's largest developing country, China faces pronounced innovation challenges in its resource-exhausted cities. Low resource pricing and costless ecological exploitation are significant factors leading to the decline of China's resource-based cities [18]. Despite a series of government documents indicating the central government's determination to promote the transformation of resource-exhausted cities, there are concerns that these support policies may merely shift local development difficulties to the central government via lacking long-term mechanisms and hindering the transformation and upgrading of resource-exhausted cities [19]. Thus, whether the support policy promotes urban innovation, establishes long-term mechanisms in market and industrial structures, and exhibits heterogeneous effects due to other factors remains unanswered and requires empirical verification.

Our study was related to three broad literatures. First, our findings connect to the literature on the transformation and development of resource-based cities. International economics literature extensively discusses the importance of the positioning of resource-based cities within global value chains and explores how to leverage international trade value chains and the smile curve theory to develop cities based on their unique resources [20–22]. These studies underscore the crucial positioning of resource-rich cities and provide recommendations for their development. Despite their lagging economic scale, resource-rich regions can achieve economic growth and industrial upgrading through global value chains [23,24]. This perspective was further validated in numerous studies. For instance, Humphrey and Schmitz (2002) indicated that by integrating into international trade value chains, resource-based cities could gain technology transfer and knowledge spillover effects, thereby promoting local industrial upgrading and economic growth [25]. This study explored the transformation of cities as their dependent resources become exhausted, particularly focusing on how to foster innovation-driven transformation in resource-exhausted cities. This is closely related to the repositioning and transformation of resource-based cities in the context of the global economy.

Second, our results are relevant to the growing body of empirical work on Chinese policy studies. China serves as a significant case study in policy research, as the government frequently and systematically experiments with different policies in various regions before deciding on nationwide implementation. These numerous pilot policies provide experimental grounds for various developmental directions in China and offer an extensive quasi-natural environment for academic research [26,27]. Regarding policies for resource-exhausted cities in China, some literature discusses its localized characteristics and impacts [28]. More closely related to our work are recent empirical studies on the effects of urban transformation policies in China. For instance, some studies examined the impact of industrial transformation and economic development efficiency [4,29], and the role of infrastructure, digital transformation, and smart city construction in promoting urban transformation and development [30–32]. Additionally, a few studies explored the development and industrial transformation of China's resource-based cities, highlighting that industrial diversification and market mechanisms are critical pathways for these cities to achieve sustainable development [33,34]. This study focused on the vulnerabilities and urgent need for transformation in resource-exhausted cities by investigating the significance of pilot transfer payment policies for their transformation and development. We further explored the mechanisms and effects of these supportive policies, enriching the existing literature on policy evaluation. Our study provides empirical evidence for understanding the effectiveness of policies in different contexts, offering valuable insights for formulating more effective policies.

Third, our findings speak to the empirical literature on the sustainability of policies. The impact of policies on the sustainable development of cities is a significant research area [35]. While supportive policies can promote innovation in resource-exhausted cities in the short term, their long-term sustainability presents challenges [36]. Our study revealed that the effects of the support policy on the innovation of resource-exhausted cities are not sustainably effective and vary across different city characteristics. The differences in policy effects may stem from variations in city-specific attributes and environmental factors [37,38]. We further analyzed the factors that influenced policy effectiveness, and thus, provided empirical insights for policy optimization. Our findings suggest that policymakers should consider the specific characteristics of resource-exhausted cities and implement more targeted policy measures to enhance long-term effectiveness and sustainability.

This study used data from 2003 to 2016 at the prefecture level to examine the impact of the support policy on the innovation of resource-exhausted cities, the differences in impact across regions and types of cities, and the long-term mechanisms of these policies. The potential contributions of this study are threefold. First, using scientific econometric methods, this study empirically analyzed the effectiveness of support policy in promoting urban innovation, supplementing theoretical research through causal inference. The quasi-natural experimental method effectively mitigated endogeneity issues in estimation, and thus, provided more reliable evidence for the transformation and upgrading of resource-exhausted cities. Second, this study was the first to analyze the long-term mechanisms of support policy in affecting urban innovation by empirically estimating the impact of marketization and industrial structure mechanisms, and thus, verified whether these policies provided long-term incentives for urban innovation. Third, using urban-level data, this study accurately identified the effects of support policy on resource-exhausted cities. Utilizing rich data across different dimensions of cities, this study deeply analyzed the heterogeneity of policy effects, thus supporting the further improvement of the support policy for resource-exhausted cities.

2. Institutional Background and Theoretical Framework

2.1. Institutional Background

The development challenges faced by China's resource-exhausted cities have garnered significant attention from the Chinese government. These cities commonly experience economic stagnation, single-industry reliance, severe environmental degradation, and underdeveloped market systems, making them weak points in China's sustainable development effort [4,39]. To promote sustainable development and transformation, the State Council issued the "Opinions on Promoting Sustainable Development in Resource-Based Cities" in 2007[1]. This document identified resource-exhausted cities as key support targets and proposed long-term mechanisms, such as resource development compensation, assistance for declining industries, resource product pricing formation, and the cultivation of alternative industries. In 2008, 2009, and 2011, the Ministry of Finance, the National Development and Reform Commission, the Ministry of Land and Resources, and the Northeast Revitalization Office of the State Council identified a total of 69 resource-exhausted cities in three batches and provided them with corresponding policy support (referred to as the support policy). In China, prefecture-level cities are administrative units under the province. Districts, counties, and county-level cities are subordinate administrative units under prefecture-level cities. Due to data availability and comparability between cities, this study concentrated on 25 pilot prefecture-level cities, with each referred to as a "city" throughout the text. The specific list of these cities is shown in Table 1. The complete list of policies can be found in Table S1 of the Supplementary Materials.

Table 1. The list of pilot cities.

First Batch (2008)	Second Batch (2009)	Third Batch (2011)
Fuxin	Fushun	Wuhai
Panjin	Qitaihe	Hegang
Liaoyuan	Huaibei	Shuangyashan
Baishan	Tongling	Xinyu
Pingxiang	Zaozhuang	Puyang
Shizuishan	Tongchuan	Shaoguan
Baiyin	Fushun	Luzhou
Jiaozuo	Jingdezhen	Zibo
Yichun		

Figure 1 reports the GDP growth trends of resource-exhausted cities compared with other cities. It is evident that the GDP growth rate of the resource-exhausted cities was significantly slower than that of other cities. Among them, the cities included in the first batch of the support policy in 2008 faced the most challenging economic development conditions. The policy proposed support for these resource-exhausted cities to promote their sustainable development. This support policy included several measures: first, establish and improve long-term mechanisms for resource development, decrease industry assistance, and resource product price formation; second, foster alternative industries through agricultural industrialization, tertiary industry development, and the development of new industries tailored to local conditions; third, address employment and social issues; fourth, environmental remediation and ecological protection; and fifth, enhance resource exploration and mining rights management. The support measures for the transformation of resource-exhausted cities were primarily implemented through transfer payments. The Ministry of Finance established transfer payments specifically for resource-exhausted cities, starting in 2007. Once selected for policy support, these cities received special transfer payments from the central government for four consecutive years. Policy documents indicate that if these cities successfully transitioned within four years, the central government would continue to provide gradually decreasing financial subsidies for an additional three years. However, in practice, even if the resource-exhausted cities succeeded in transitioning or reached the end of the funding period, the central government continued to support them through special transfer payments. By 2018, the cumulative support amount had reached nearly CNY 160 billion[2]. Additionally, the policy not only provided special financial support but also clarified the use of funds and establishes a transfer payment system. To further enhance the effectiveness of financial assistance, provincial governments also introduced supporting policies, special funds, and sustainable development loans. These special financial funds were used to develop alternative industries, primarily by supporting industrial development projects, building industrial parks, and constructing infrastructure.

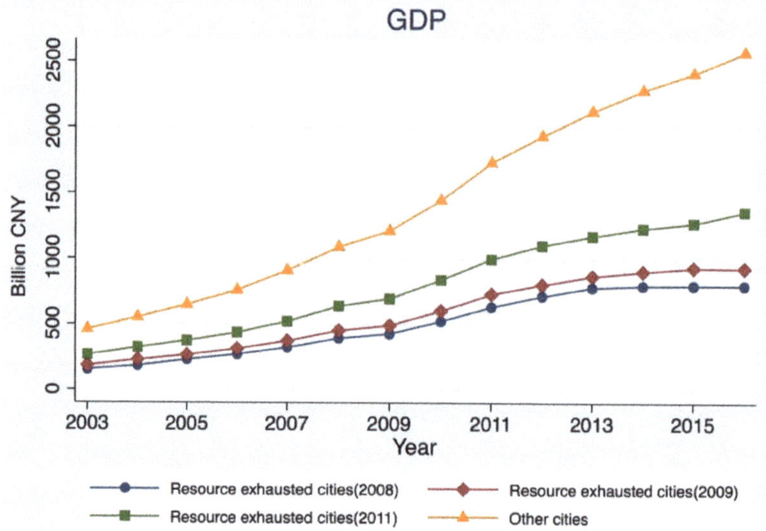

Figure 1. GDP growth trends in Chinese cities.

2.2. Hypothesis Development

Despite the central government's provision of special transfer payments and supporting policy guidance to address the sustainable development issues of resource-exhausted cities, whether these support policies effectively promote innovation in these cities still requires further scientific examination. In particular, it is necessary to investigate whether they have fostered market-oriented reforms and the establishment of long-term mechanisms, such as industrial structure upgrades.

On one hand, the low level of marketization in resource-exhausted cities constrains their development [4]. Influenced by China's early planned economy, resource-based industries have implemented highly centralized planned resource allocation methods, resulting in long-term monopolistic conditions in resource-exhausted cities and low market vitality [40]. The low pricing of resources not only fails to reflect the true supply and demand relationships in the market but also weakens the market system in resource-exhausted cities [39,41]. This can lead to rent-seeking issues, distort the market allocation of factors, and negatively impact the availability of innovation elements in these cities. In response, the support policy explicitly proposed "fully leveraging the fundamental role of the market in resource allocation and stimulating the intrinsic vitality of various market entities". An increase in marketization can significantly promote urban innovation. Higher marketization implies improved overall resource allocation efficiency in cities, with innovative elements shifting from low-efficiency to high-efficiency sectors, thereby promoting urban innovation [42]. Additionally, higher marketization means reduced government intervention in the market, creating a fairer and more orderly market environment [43]. This enhances competition as private enterprises grow, compelling local enterprises to innovate to stay competitive, thereby boosting urban innovation [44]. Based on this analysis, we propose the following research hypothesis:

H1. *By increasing the level of marketization, the support policy for resource-exhausted cities promotes urban innovation.*

On the other hand, the problem of a single industrial structure is prominent in resource-exhausted cities [34]. Early development in these cities centered around resource-based industries, with other industries primarily serving the resource sector. Thus, resource depletion also leads to the decline of other industries in the city [45]. The key to the transformation and development of resource-exhausted cities lies in industrial transformation, which drives sustainable economic development through industrial upgrading [4,46]. The support policy clearly states the need to "establish industrial assistance mechanisms" and "accelerate the development of the tertiary sector". Therefore, the support policy may drive the industrial structure upgrade in resource-exhausted cities, which, in turn, promotes urban innovation, as confirmed by numerous studies [47–49]. Industrial structure upgrades can facilitate technology diffusion and knowledge spillovers within and between cities, increasing the demand for innovation and directly promoting autonomous urban innovation [50,51]. Moreover, industrial structure upgrades can expand niche markets, stimulate demand growth, intensify market competition, and encourage enterprises to innovate to enhance competitiveness, thereby driving the overall urban innovation [52,53]. Based on the above, we propose the following research hypothesis:

H2. *By upgrading the industrial structure, the support policy for resource-exhausted cities promotes urban innovation.*

3. Methodology and Data

3.1. Data Source

The Urban Innovation Data originate from the China Research Data Service Platform (CNRDS). The CNRDS is a high-quality, specialized platform for Chinese innovation research data. The Chinese Innovation Research Database (CIRD) is part of this platform and was developed based on patent applications and authorizations from various listed and non-listed companies, as well as different regions. The data quality is reliable and was utilized in other high-quality research studies [54,55]. This study used the number of city patent authorizations as an indicator of urban innovation [38,56,57].

Other city characteristic variables primarily come from the "China City Statistical Yearbook" and statistical bulletins of various cities. Following the methodology of Brandt et al., 2006 and 2012 [58,59], these variables were deflated using the year 2000 as the base year. These data were utilized for control variables, mechanism analysis, and other heterogeneity analyses.

To comprehensively reflect the changes in resource-exhausted cities before and after the policy implementation and accurately measure the effects of the policy, this study selected data from five years before the pilot began and five years after the pilot ended, spanning 2003–2016. Given the significant differences in economic development levels and administrative status between the four direct-controlled municipalities (Beijing, Tianjin, Shanghai, and Chongqing) and the Hong Kong, Macau, and Taiwan regions compared with prefecture-level cities, this study excluded samples from these four municipalities and three regions. First, since these cities are directly governed by the central government and are equivalent to provincial-level units in terms of administrative and political status, with significant differences in GDP and political status, this makes direct comparisons inappropriate. In related literature, these cities are often excluded or treated separately. Second, the data for some cities, such as Macau and Hong Kong, as well as Taiwan, are not available, making it impossible to conduct a comprehensive and accurate analysis. Additionally, samples with missing key variables are excluded. Consequently, the final dataset of this study included 3878 samples from 277 prefecture-level cities over 14 years, with the treatment group comprising 25 prefecture-level cities. The definitions of variables and descriptive statistics are presented in Table 2.

Table 2. Variable definitions and descriptive statistics.

Panel A. Variable Definitions	
Variable	**Definition**
CityInnov	Log value of number of patents granted
Control Variables	
MineRatio	Proportion of mining industry in the total output
FiscalExp	Log value of government expenditure
PCW	Log value of per capita wage
UnempRate	Unemployment rate
PCTRS	Log value of per capita retail sales of consumer goods
Mechanism Variables	
Industry	Ratio of the value added of the tertiary industry to the total GDP
Market	Fan Gang's marketization index

Panel B. Descriptive Statistics					
Variable	**Obs**	**Mean**	**SD**	**Min**	**Max**
CityInnov	3878	6.025	1.770	1.099	11.419
MineRatio	3878	0.107	0.169	0	0.793
FiscalExp	3878	13.633	0.940	10.410	17.219
PCW	3878	9.932	0.491	2.225	11.675
UnempRate	3878	3.349	0.770	1.310	5.060
PCTRS	3878	8.673	0.865	−1.337	11.580
Industry	3878	0.364	0.084	0.086	0.764
Market	3878	8.893	2.678	1.959	16.949

3.2. Empirical Methodology

The Chinese government announced the list of resource-exhausted cities in 2008, 2009, and 2011, which provided an opportunity to employ a "quasi-natural experiment" approach. Within this institutional context, three batches of cities in the treatment group received interventions by following a typical staggered difference-in-differences (DID) framework. First, the DID model effectively controlled for time and individual fixed effects, which reduced the bias from omitted variables. Second, the DID model was particularly suitable for evaluating the differences between the affected and unaffected groups before and after the policy implementation, which aligned with our research on the impact of the support policy on urban innovation in resource-exhausted cities. Therefore, referencing Beck et al., 2010 [60], we constructed a staggered DID model (Equation (1)) to determine the causal impact of the support policy for resource-exhausted cities on urban innovation:

$$CityInnov_{i,t} = \alpha + \beta Traet_i \times Post_{it} + \varphi X'_{it} + \lambda_i + \lambda_t + \lambda_i t + \varepsilon_{it} \quad (1)$$

where i and t represent the city and year, respectively. $CityInnov_{i,t}$ represents the natural logarithm of the total number of patent grants in city i at year t. $Traet_i$ is a dummy variable representing whether a city belonged to the treatment group, with resource-exhausted cities assigned a value of 1, and others assigned a value of 0. $Post_{it}$ is used to identify whether a city has been affected by the policy. If a city was listed as a resource-exhausted city in a given year, it took a value of 1; otherwise, it was 0. X'_{it} is a vector of control variables composed of various city characteristics, the specifics of which are elaborated upon below. λ_i represents city fixed effects, and λ_t denotes time fixed effects. We included $\lambda_i t$ fixed effects to control for development trends of different cities across different years [61]. ε_{it} represents city-clustered robust standard errors to mitigate potential inter-group correlation issues [62,63].

The feature vector X_{it} is the interaction of the predetermined variables with the time dummy variables λ_t, i.e., $Z_{i2007} \times \lambda_t$ [64]. This was to control for factors that influenced the selection of pilot cities and eliminate the possibility that these factors could change over time and affect urban innovation.

The predetermined variable Z_{i2007}, which refers to a series of control variables influencing the selection of policy pilots, was primarily chosen based on the standards and procedures for defining resource-exhausted cities. The National Development and Reform Commission, the Ministry of Natural Resources, and the Ministry of Finance jointly established the principles, standards, and work plan for defining resource-exhausted cities. This followed a declaration-based principle, with applications submitted by provincial governments. The final list was determined based on a comprehensive scoring system that included four major categories of indicators: resource reserves, mining development, people's livelihood, and fiscal economy. Even if a city met the scoring standards, it may not be included in the list for funding if the provincial government did not apply or due to intense competition, as seen in the case of Jixi City[3]. This means that controlling for the four predetermined conditions, the selection of resource-exhausted cities was more random. Additionally, resource-exhausted cities were characterized by a declining industrial efficiency, a single industrial structure, weak local financial strength, and low worker incomes [4,39,65]. Based on the data availability, this study selected several predetermined variables: the proportion of extractive industries (MineRatio), the number of state-owned enterprises (SOEs), fiscal expenditure (FiscalExp), fiscal deficit ratio (DeficitRatio), average wage (PCW), unemployment rate (UnempRate), and per capita retail sales of consumer goods (PCTRS). This study used data from 2007, which was the year before the first list of resource-exhausted cities was announced, to verify through regression analysis whether these variables influenced a city's inclusion in the list of resource-exhausted cities. Table 3 shows the results of an employed stepwise regression method for verification, which revealed that the above variables significantly impacted whether a city was listed as a resource-exhausted city. Therefore, the vector of predetermined variables Z_{i2007} that influenced the selection of resource-exhausted cities should include the proportion of extractive industries (MineRatio), fiscal expenditure (FiscalExp), average wage (PCW), unemployment rate (UnempRate), and per capita retail sales of consumer goods (PCTRS).

Table 3. Pre-treatment analysis.

	(1) $Treat_i$	(2) $Treat_i$	(3) $Treat_i$
MineRatio	0.360 ***	0.337 ***	0.374 ***
	(0.099)	(0.099)	(0.097)
SOEs	−0.069 ***	−0.014	−0.016
	(0.024)	(0.034)	(0.031)
FiscalExp		−0.107 ***	−0.120 ***
		(0.037)	(0.042)
DeficitRatio		−0.570 *	0.032
		(0.332)	(0.453)
PCW			−0.204 **
			(0.095)
UnempRate			10.926 ***
			(3.831)
PCTRS			0.110 **
			(0.049)
Observations	277	277	277
R-squared	0.087	0.118	0.186

Note: ***, **, and * represent the significant levels of 1%, 5%, and 10%, respectively. Standard errors are in parentheses.

4. Empirical Results

4.1. Baseline Results

The baseline regression results are presented in Table 4. Column (1) shows the regression results without any control variables or fixed effects, with a regression coefficient of 1.143 and an R^2 of 0.051. Column (2) shows the estimation results with λ_i, λ_t, and $\lambda_i t$ city and year fixed effects, as well as city fixed effects interacted with year. In this case, the coefficient decreased to 0.216, but due to the inclusion of fixed effects, the R^2 increased to 0.979. Column (3) further incorporates the vector of city characteristics. From the regression results in each column, it can be observed that the support policy for resource-exhausted cities significantly increased the number of patent grants in pilot cities, indicating a positive impact of these policies on innovation in resource-exhausted cities. Column (3) shows that the regression coefficient for the core independent variable was 0.249, and it was significant at the 1% level. This indicates that the support policy, through transfer payments and corresponding institutional guidance, increased the logarithm of the number of patent grants in resource-exhausted cities by 0.249. This coefficient had important economic implications, signifying that the support policy increased the number of patent grants in resource-exhausted cities by an average of 28.44% compared with the pre-policy average number of patent grants ($\left(e^{\ln(167.32+1)+0.249} - 1 - 167.32\right)/167.32$). The support policy for resource-exhausted cities significantly enhanced the level of urban innovation.

Table 4. The average effect of policy assistance on urban innovation.

	(1) CityInnov	(2) CityInnov	(3) CityInnov
$Treat_i \times Post_{it}$	1.143 ***	0.216 **	0.249 ***
	(0.120)	(0.093)	(0.091)
λ_i	No	Yes	Yes
λ_t	No	Yes	Yes
$\lambda_i t$	No	Yes	Yes
X'_{it}	No	No	Yes
Observations	3878	3878	3878
R-squared	0.051	0.979	0.981

Note: *** and ** represent the significant levels of 1% and 5% respectively. Standard errors are in parentheses.

4.2. Parallel Trend Test

The most important assumption of the difference-in-differences (DID) model is the comparability between the treatment group and the control group, also known as the parallel trend assumption. We used an event study approach to test whether the number of authorized patents in pilot cities and non-pilot cities had the same trend before policy implementation. Following Beck et al., 2010 [60], we constructed Equation (2):

$$CityInnov_{i,t} = \alpha + \sum_{k=-5}^{k=+5} \beta_k Treat_i \times Year_k + X'_{it}\gamma + \lambda_i + \lambda_t + \lambda_i t + \varepsilon_{it} \tag{2}$$

In this model, $Year_k$ is a series of dummy variables that equal 1 when city i is a resource-exhausted city and is in the $Year_k$ year of policy implementation, and 0 otherwise. All other terms have the same meanings as in Equation (1). The regression coefficients β_k indicates whether there is a significant difference in the number of patent grants between the treatment and control groups in the kth year of the policy implementation. The period before policy implementation was set as the baseline (reference period), and we trimmed the extreme values of the observations to examine the dynamic changes over the five years before and after the policy implementation.

Figure 2 visualizes the regression results and reports the 95% confidence intervals. It can be observed that when $k < 0$, the estimated coefficient β_k was not significantly different from zero, indicating that before the implementation of the support policy for resource-exhausted cities, there was no significant difference in the trend of patent grant numbers between the treatment and control groups. This confirmed the parallel trend assumption. The dynamic effects show that the support policy for resource-exhausted cities had a significant positive impact on urban innovation. On the one hand, the policy had a certain lag effect, where it became significantly positive in the second year after implementation. On the other hand, as time progressed, this innovation-promoting effect began to weaken, and by the fifth year after the policy implementation, the estimated coefficient β_k was not significant, indicating poor policy sustainability.

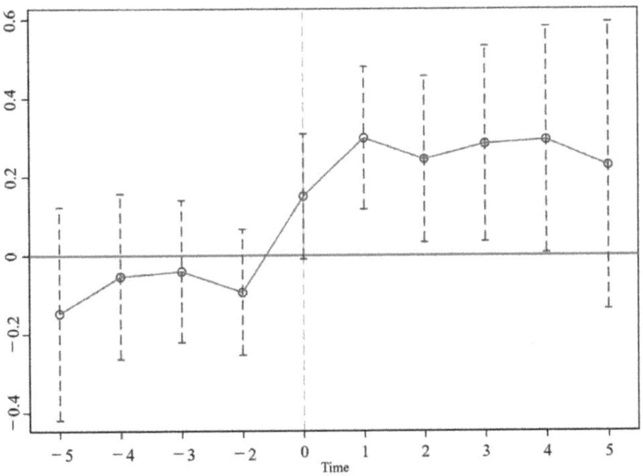

Figure 2. Dynamic effect on urban innovation.

4.3. Robustness Checks

Building on the baseline regression, this study more cautiously verified the robustness of the promotion effect of the support policy for resource-exhausted cities on urban innovation. We conducted robustness checks by using balance tests, replacing the dependent variable, altering the representation of policy implementation time, shortening the window period, substituting the control group, and performing placebo tests.

4.3.1. Balance Test

Differences between the treatment and control groups before policy implementation are a significant cause of estimation bias in baseline results [66]. Therefore, considering the control variables, we examined other factors that might influence urban innovation before policy implementation to verify the sample balance. The econometric model we used is shown in Equation (3):

$$UCV_{i2007} = \alpha + \beta Treat_i + \gamma Z_{i2007} + \varepsilon_i \tag{3}$$

where UCV_{i2007} represents the observable other urban characteristic variables of cities in 2007. We selected six variables: GDP (GDP), fiscal deficit (Deficit), number of state-owned enterprises (SOEs), expenditure on science and education (ESTE), number of higher education institutions (College), and number of scientific personnel (Tec_Prac). If the estimated coefficient β is significant, it indicates heterogeneity between the two groups of samples. Table 5 reports the estimation results of Equation (3). After including the control variables, the estimated coefficients of the aforementioned variables on β are all insignificant, indicating that the samples pass the balance test and that the treatment and control groups are homogeneous and comparable.

Table 5. The balance test.

	(1) GDP	(2) Deficit	(3) SOEs	(4) ESTE	(5) College	(6) Tec_Prac
Treat_i	−0.026	0.030	−0.058	−0.060	0.030	−0.122
	(0.060)	(0.103)	(0.113)	(0.045)	(0.134)	(0.153)
Z_{i2007}	Yes	Yes	Yes	Yes	Yes	Yes
Observations	277	277	277	277	277	277
R-squared	0.912	0.593	0.541	0.920	0.564	0.440

Note: Standard errors are in parentheses.

4.3.2. Replace the Variable Measurement

To further validate the robustness of our results, we replaced the measurement methods of the dependent and explanatory variables. For the dependent variable, we utilized the Urban Innovation Index released by Fudan University in the "China City and Industrial Innovation Report 2017" to measure the urban innovation levels by replacing the original dependent variable [67]. This index, like the number of authorized patents used in this study, served as an indicator of urban innovation capacity. However, the measurement method differed. The Urban Innovation Index estimates the average value of patents based on legal status updates and annual fee structures, and comprehensively considers a city's innovation output and entrepreneurial activities, and thus, represents the city's innovation level [68,69]. The regression results are reported in columns (1) and (2) of Table 6, where the regression coefficient of the core independent variable remained significantly positive. This indicates that the support policy for resource-exhausted cities significantly increased the urban innovation index of pilot cities, and thus, it had a positive impact on urban innovation.

Table 6. The effect of policy assistance on urban innovation: robustness 1.

	(1)	(2)	(3)	(4)	(5)
	InnovIndex			CityInnov	
$\text{Treat}_\text{i} \times \text{Post}_\text{it}$	2.209 ***	1.730 ***		0.227 ***	0.234 ***
	(0.649)	(0.546)		(0.087)	(0.090)
$\text{Treat}_\text{i} \times \text{Post}_\text{it}'$			0.232 **		
			(0.093)		
λ_i	Yes	Yes	Yes	Yes	Yes
λ_t	Yes	Yes	Yes	Yes	Yes
$\lambda_\text{i}\text{t}$	Yes	Yes	Yes	Yes	Yes
X_it	No	Yes	Yes	Yes	Yes
Observations	3822	3822	3878	3324	3047
R-squared	0.937	0.938	0.981	0.983	0.983

Note: *** and ** represent the significant levels of 1% and 5% respectively. Robust standard errors clustered at the city level are in parentheses.

Regarding the independent variables, considering the differences in the announcement timings of the three batches of resource-depleted cities [70], the variable $Traet_i \times Post_{it}$ was set based on the announcement months of each batch separately. Specifically, the specific announcement dates for the three batches of resource-depleted cities were 17 March 2008, 5 March 2009, and 15 November 2011. Therefore, for each year when the resource-depleted cities were announced, the respective values were set to 3/4, 5/6, and 1/12, while the values for other years remained unchanged. The estimation results are reported in column (3) of Table 6, where the estimated coefficient of the core independent variable remained positive and significant at the 5% level, which enhanced the robustness of the baseline regression results.

4.3.3. Shortening the Window Period

The choice of the window period may affect the estimation results. This study utilized a sample spanning from 2003 to 2016, which totaled 14 years. One potential concern was that the regression results might be driven by the relatively long window period. To verify the robustness of the baseline regression results under a reduced sample size and shortened window periods, we conducted regressions using samples from 2004 to 2015 and 2005 to 2015, where the sample window was gradually narrowed. The results are shown in columns (4) and (5) of Table 6. It can be observed that there were no significant differences between the core explanatory variables and baseline regression results. The regression coefficients remained significantly positive, which further enhanced the robustness of the baseline regression results.

4.3.4. Changing the Control Group

Considering the crucial importance of comparability between the experimental and control groups in the DID method, we set multiple control groups to verify the robustness of the baseline results.

(1) Resource-based cities: Given potential differences between resource-based cities and general cities, we only selected resource-based cities as control groups. The list of resource-based cities was sourced from China's "National Sustainable Development Plan for Resource-Based Cities (2013–2020)" published in 2013, which listed 126 prefecture-level resource-based cities and covered all samples of resource-exhausted cities. We used this published list of resource-based cities as the control groups for the regression. The regression results are shown in column (1) of Table 7 and indicate that the impact of resource-exhausted city support policy on urban innovation remained significantly positive.

Table 7. The effect of policy assistance on urban innovation: robustness 2.

Control Groups	(1) Resource-Based Cities CityInnov	(2) The Same Province CityInnov	(3) PSM-DID CityInnov
$Treat_i \times Post_{it}$	0.199 **	0.244 ***	0.241 ***
	(0.088)	(0.093)	(0.092)
λ_i	Yes	Yes	Yes
λ_t	Yes	Yes	Yes
$\lambda_i t$	Yes	Yes	Yes
X_{it}	Yes	Yes	Yes
Observations	1582	2492	3535
R-squared	0.969	0.979	0.982

Note: *** and ** represent the significant levels of 1% and 5% respectively. Robust standard errors clustered at the city level are in parentheses.

(2) Cities in the same province: Considering that cities in the same province as resource-exhausted cities have more similar political, economic, and transportation conditions, we chose cities in the same province as resource-exhausted cities as control groups to enhance the comparability. After retaining samples that contained resource-exhausted cities within the province, the regression results are shown in column (2) of Table 7, where the regression coefficient of the core independent variable was still significantly positive.

(3) PSM-DID: Using the propensity score matching (PSM) method, we adopted a 1:1 nearest neighbor caliper matching method to rematch the pilot samples with more similar control groups and then re-run the regression using the matched samples. Figure 3 displays the kernel density distribution of the propensity scores before and after matching between the treatment and control groups. It can be observed that after matching, the difference in kernel density distribution between the two groups significantly decreased, indicating that the PSM significantly enhanced the comparability between the experimental and control groups. Column (3) of Table 7 presents the estimation results of PSM-DID, where the regression coefficient of the core independent variable remained significant at

the 1% significance level. A series of tests that replaced the control groups all strengthened the robustness of the baseline regression results.

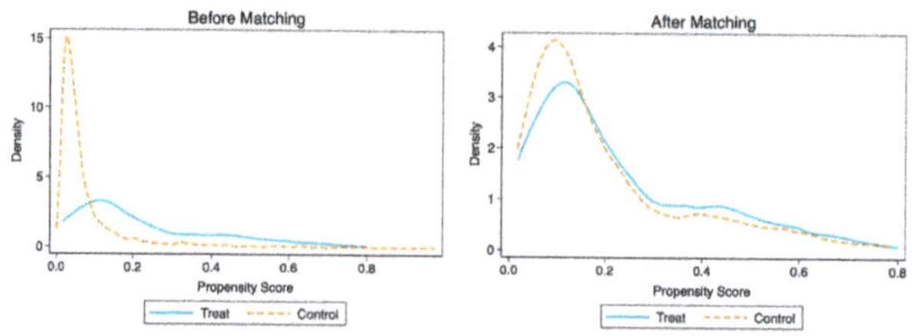

Figure 3. Kernel density of the treatment and control groups.

4.3.5. Placebo Tests

Despite incorporating multiple control variables and fixed effects in the baseline model, concerns persisted that the effects of the resource-exhausted city support policy may stem from other unobservable factors. This study randomly generated policy implementation times and treatment group samples as placebos [71,72]. Specifically, we randomly assigned policy effective times to each city based on the policy implementation process and non-repetitively selected nine, eight, and eight cities in three batches from all cities to be added to the virtual list of resource-exhausted cities as experimental groups for regression. We repeated this randomization process 500 times. The kernel density distributions of all estimated coefficients and *p*-values are shown in Figure 4, which indicates that the estimated coefficients of the placebo variables were closely associated with a normal distribution that was primarily centered around zero and lacked statistical significance. In contrast, the estimated values obtained from the baseline regression were distinguished by the red dashed line and displayed significant deviations from zero. Therefore, the results of the placebo tests reinforced the conclusion drawn from the baseline results that the effect of urban innovation promotion stemmed from the resource-exhausted city support policy rather than other unobservable random factors.

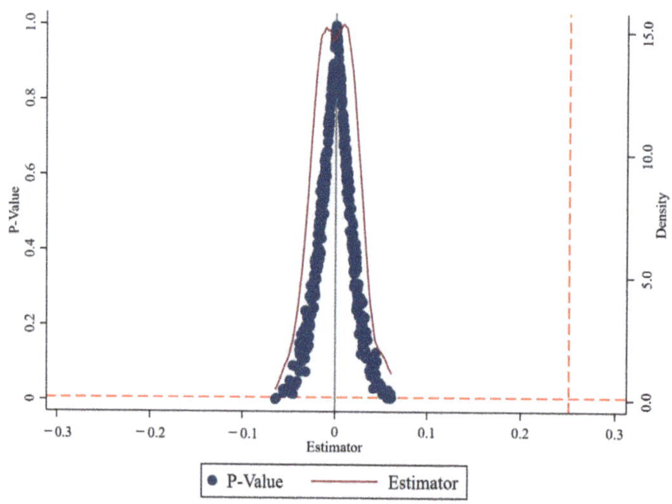

Figure 4. Placebo test. Note: The red horizontal dashed line represents p = 0.007, and the red vertical dashed line represents $\beta = 0.249$, which are the results of the baseline regression.

4.4. Mechanism Analyses

Based on the background and theoretical analysis, we proposed two mechanistic hypotheses to explain how supportive policies impacted the innovation in resource-exhausted cities: the degree of marketization and the industrial structure of the city. In this section, we describe the two-step method used to validate these hypotheses [73,74]. The specific model was as follows:

$$MV_{i,t} = \alpha + mTraet_i \times Post_{it} + \varphi X'_{it} + \lambda_i + \lambda_t + \lambda_i t + \varepsilon_{it} \tag{4}$$

The two-step method for the mechanism testing primarily included two equations, namely, Equations (1) and (4). Equation (1) was already validated in the previous section. In Equation (4), $MV_{i,t}$ represents the mechanism variables *Market* and *Industry*, and the meanings of the other terms remain consistent with those in Equation (1). We aimed to verify the effect of the regression coefficient m of the policy on the mechanism variables. The detailed analysis of the two mechanisms is provided below.

4.4.1. Marketization

The marketization level is one of the important factors influencing regional innovation levels. In resource-exhausted cities, many enterprises are resource-dependent state-owned enterprises, and the phenomenon of integration between the government and enterprises is severe. High market entry barriers for private enterprises result in generally low levels of marketization in these cities. The policy documents explicitly state the need to fully utilize the market to allocate resources and improve the efficiency of resource-exhausted cities. Therefore, this study explored the mechanism by which the resource-exhausted city support policy promoted urban innovation by examining whether they enhanced the city's marketization level. Drawing on the calculation method of Fan Gang's marketization index [75], we calculated the urban marketization index to measure the level of marketization.

Table 8 reports the estimated results of the impact of resource-exhausted city support policy on urban marketization. Column (1) presents results without controlling for any variables or fixed effects (FEs), with a coefficient of 3.233, which was significant at the 1% level. As shown in column (2), after adding fixed effects, the coefficient decreased to 0.152, significant at the 5% level. It can be observed in the third column that the net impact of the policy on urban marketization was 0.186, which was significant at the 5% level. This indicates that the support policy significantly improved the marketization level of resource-exhausted cities, which enhanced the efficiency of market factor allocation and thereby promoted urban innovation [76]. The research hypothesis H1 of this study was validated.

Table 8. Results of mechanism analysis: marketization.

	(1) Marketization	(2) Marketization	(3) Marketization
$Treat_i \times Post_{it}$	3.233 ***	0.152 **	0.186 **
	(0.161)	(0.070)	(0.074)
λ_i	No	Yes	Yes
λ_t	No	Yes	Yes
$\lambda_i t$	No	Yes	Yes
X'_{it}	No	No	Yes
Observations	3878	3878	3878
R-squared	0.0669	0.990	0.990

Note: *** and ** represent the significant levels of 1% and 5%. Robust standard errors clustered at the city level are in parentheses.

4.4.2. Industrial Structure

The single structure of industries is another factor constraining innovation in resource-exhausted cities. With resource-dependent industries at their core, resource-exhausted cities face severe impacts on their development as resources become depleted. The resource-exhausted city support policy proposed measures such as establishing industrial assistance mechanisms and vigorously developing alternative industries to promote the upgrade of urban industrial structure and achieve sustainable development. This study used the ratio of the value added of the tertiary industry to the total GDP to measure industrial structure upgrades.

Table 9 reports the estimated results of the impact of resource-exhausted city support policy on the urban industrial structure. As shown in column (1), when not controlling for pre-determined variables and any FEs, the estimated coefficient of the policy was very small and not significant. As shown in column (2), after adding FEs, the estimated coefficient was 0.022 at the 1% level. In the most stringent model control, as shown in column (3), which controlled for pre-determined variables and FEs, the estimated coefficient of the policy was 0.013 and still significant at the 10% level. This indicates that the resource-exhausted city support policy significantly increased the ratio of the tertiary industry to the secondary industry in urban areas, which promoted industrial structure upgrades. Industrial upgrades, by fostering the spillover of technology and knowledge within cities and increasing market competitiveness, were shown in the literature to effectively promote urban innovation [77]. Thus, hypothesis H2 of this study was confirmed.

Table 9. Results of mechanism analysis results: industrial structure.

	(1) Industrial Structure	(2) Industrial Structure	(3) Industrial Structure
$Treat_i \times Post_{it}$	0.003	0.022 ***	0.013 *
	(0.009)	(0.007)	(0.007)
λ_i	Yes	Yes	Yes
λ_t	Yes	Yes	Yes
$\lambda_i t$	Yes	Yes	Yes
X'_{it}	No	No	Yes
Observations	3878	3878	3862
R-squared	0.0276	0.942	0.949

Note: *** and * represent the significant levels of 1% and 10% respectively. Robust standard errors clustered at the city level are in parentheses.

4.5. Heterogeneity Analyses

The above results indicate that the support policy effectively enhanced the innovation performance of resource-exhausted cities. Considering the differences between cities, this policy may have different effects on resource-exhausted cities with different characteristics. Therefore, this study analyzed the heterogeneity of policy effects in different regions and types of resource-exhausted cities.

4.5.1. Region

China is vast in territory, with significant differences in economic, transportation, and natural environments across regions. The National Bureau of Statistics of China divides provinces into eastern, central, western, and northeastern regions, with the eastern region being geographically and economically superior to other regions [78]. Therefore, we divided the sample into eastern and non-eastern regions for the regression analysis. The regression results are reported in columns (1) and (2) of Table 10. It can be observed that whether in the eastern region or the non-eastern region, the support policy had a significant promoting effect on the innovation level of resource-exhausted cities. However, the effect of promoting urban innovation in the eastern region was more significant. This may have been because, compared with other regions, the eastern region was more economically developed, with better infrastructure and a greater concentration of innovative elements [29,79]. These objective advantages provided a better innovation environment for resource-exhausted cities in the eastern region, which made the promotion effect of the support policy on innovation more significant in the eastern region.

Table 10. Results of heterogeneity analyses: region, types of resource, and mining dependency.

	Region		Types of Resource		Mining Dependency	
Variables	(1) Eastern CityInnov	(2) Others CityInnov	(3) Coal Resources CityInnov	(4) Other Resources CityInnov	(5) Low CityInnov	(6) High CityInnov
$Treat_i \times Post_{it}$	0.379 *** (0.084)	0.238 ** (0.105)	0.136 (0.166)	0.274 *** (0.101)	0.558 ** (0.270)	0.168 * (0.092)
λ_i	Yes	Yes	Yes	Yes	Yes	Yes
λ_t	Yes	Yes	Yes	Yes	Yes	Yes
$\lambda_i t$	Yes	Yes	Yes	Yes	Yes	Yes
X_{it}	Yes	Yes	Yes	Yes	Yes	Yes
Observations	1162	2716	406	1176	1946	1932
R-squared	0.984	0.974	0.972	0.971	0.984	0.975

Note: ***, **, and * represent the significant levels of 1%, 5%, and 10%, respectively. Robust standard errors clustered at the city level are in parentheses.

4.5.2. Types of Resources

Each resource-type city harbors different types of resources, and the objective differences in resource endowments prompt the differentiated development of urban industrial structures, which may affect the implementation of the support policy. The Development Research Center of the State Council of China categorizes resource-type cities into five types: coal, forest industry, metallurgy, petroleum, and others. At the prefectural level, coal-type cities account for a large proportion of resource-exhausted cities. Additionally, due to the greater difficulty in the industrial restructuring of coal-type cities, they are the key targets of policy support. Therefore, we divided the sample of resource-type cities into coal-type cities and other cities for separate analysis. The estimation results are reported in columns (3) and (4) of Table 10. We found that the impact of the support policy on coal-type cities was not statistically significant, while the promoting effect of innovation levels in other types of resource-exhausted cities was significant at the 1% level. This aligned with our expectations; as compared with the coal industry, the supporting industries in the petroleum and metallurgical industries were more technologically advanced and had more complete industrial chains, which were more conducive to urban innovation [80]. The results suggest the need for greater attention to the transformation and innovation of coal-type cities.

4.6. *Further Analyses*

This study further explored the factors that influenced the effectiveness of the support policy, with the aim to understand how differences in urban characteristics affect policy implementation effectiveness and provide insights for policy optimization.

4.6.1. Mining Dependency

The higher the proportion of the mining industry in the total industrial output value of a city, the lower the level of manufacturing development, which implies greater challenges for innovation assistance in such resource-exhausted cities [81]. Since it is difficult to obtain industry output data at the city level, we followed the approach of Sun and Abraham [82] and aggregated micro-enterprise output data from the China Industrial Enterprise Database to obtain data on the proportion of the mining industry at the city level. During the integration process, samples with liquid assets exceeding total assets, fixed assets exceeding total assets, missing enterprise codes, and incorrect establishment years were excluded [59,83]. This study conducted group regression based on the median proportion of the mining industry in 2007. The regression results in columns (5) and (6) of Table 10 indicate that in cities with a higher dependence on the mining industry, the impact of the support policy on urban innovation was smaller and less significant. This suggests that differences in urban industrial foundations affected the urban innovation promotion effect of the support policy for resource-exhausted cities. Overreliance on the mining industry in resource-exhausted cities exacerbated the problem of industrial structure singularity, further limiting the promotion effect of the support policy on urban innovation. Therefore, policymakers need to consider further guiding these cities to move away from resource dependence and transform their development dynamics.

4.6.2. Talent Environment

Innovation can be seen as an output process of input factors, where the input of human resources has a crucial impact on innovation output [84]. This study explored whether the regional talent environment affected the innovation promotion effect of the support policy for resource-exhausted cities, with a focus on two aspects: the number of higher education institutions and the number of science and technology professionals. On one hand, higher education institutions are one of the key drivers of regional innovation, with advantages in fundamental research, abundant scientific talents, and interdisciplinary integration [50]. We divided the samples into two groups based on the median number of universities in the cities and conducted regression analyses separately.

The regression results in columns (1) and (2) of Table 11 indicate that resource-exhausted cities with a higher number of universities showed a more significant improvement in urban innovation after being influenced by the support policy. On the other hand, talent was the primary resource for innovation and the foundational element for innovative development [85]. Therefore, a higher number of science and technology professionals were conducive to promoting urban innovation. Since only four resource-exhausted cities had a number of science and technology professionals above the median in the total sample, we grouped the samples based on the median number of science and technology professionals in resource-exhausted cities and conducted regression analyses. The results in columns (3) and (4) of Table 11 similarly demonstrate that the support policy had a stronger and more significant innovation promotion effect on resource-exhausted cities with more science and technology professionals. In other words, the issue of talent drain in resource-exhausted cities significantly limited the innovation promotion effect of the support policy. Addressing the supply of human resources in the transformation and development of resource-exhausted cities is a problem that policies need to further solve.

Table 11. Results of heterogeneity analyses: talent environment.

	Higher Education Institutions		Science and Technology Professionals	
Variables	**(1) Low Quantity**	**(2) High Quantity**	**(3) Low Quantity**	**(4) High Quantity**
$Treat_i \times Post_{it}$	0.169	0.393 **	0.212	0.172 **
	(0.116)	(0.164)	(0.150)	(0.077)
λ_i	Yes	Yes	Yes	Yes
λ_t	Yes	Yes	Yes	Yes
$\lambda_i t$	Yes	Yes	Yes	Yes
X_{it}	Yes	Yes	Yes	Yes
Observations	1708	2170	770	3108
R-squared	0.965	0.984	0.958	0.984

Note: ** represent the significant levels of 5%. Robust standard errors clustered at the city level are in parentheses.

5. Conclusions

The sustainable development transformation of resource-exhausted cities is a global challenge. Urban innovation is a key pathway to achieving sustainable urban development, making it crucial to explore ways to enhance the innovation capacity of resource-exhausted cities, both theoretically and practically. This study utilized the exogenous shock from China's support policy to construct a difference-in-differences model, which empirically analyzed the impact of these policies on urban innovation and their long-term mechanisms based on panel data from 277 prefecture-level cities in China from 2003 to 2016. The main conclusions of this study are summarized as follows: First, the support policy for resource-exhausted cities significantly promotes innovation. Specifically, the implementation of central transfer payments and supporting policies increased the number of authorized patents by 28.34%. Second, our findings indicate that the support policy established two long-term mechanisms—industrial structure upgrades and increased marketization—which substantially enhanced the urban innovation levels. Third, the effect of these policies on urban innovation varies depended on the region, city resource type, reliance on extractive industries, and regional talent supply. Specifically, the innovation-promoting effect of the policies was more pronounced in cities located in the eastern region, non-coal cities, cities with low reliance on extractive industries, and cities with higher levels of science and education.

In this critical era of global sustainable development, this study used empirical data to investigate the impact of government support policy on the innovation of resource-exhausted cities, which has significant policy implications. A fundamental recommendation is that the economic transformation of resource-exhausted cities requires corresponding financial and institutional support from the government, alongside differentiated support policies. These cities face poor economic foundations, insufficient growth momentum, and severe talent loss, making it difficult to accelerate urban innovation based on their own resources, thereby achieving sustainable development transformation. According to the main findings of this study, the government should consistently provide financial support and guidance on transformation systems to promote urban innovation and sustainable development.

Second, based on the results of mechanism analysis, policies should enhance the transformation of resource-exhausted cities from two aspects: industrial structure and market orientation. This study found that upgrading the industrial structure is pivotal in promoting innovation in resource-exhausted cities through supportive policies. To advance this transformation, it is essential to extend dominant industries, develop alternative sectors, and foster a diversified industrial system that promotes growth in both established and emerging industries, thereby driving urban innovation transformation. Moreover, the traditional government-led economy in resource-exhausted cities significantly stifles innovation dynamics. Therefore, accelerating market-oriented reforms in these cities to foster urban innovation is crucial. On one hand, the government should streamline administration, decentralize authority, reduce economic interventions, and allow the market to play a decisive role in resource allocation to create an efficient market environment. On the other hand, enhancing market supervision through improved legal frameworks to safeguard innovation outcomes and actively guiding urban transformation is essential.

Third, based on further research findings, it is evident that the support policy had varying effects on different types of resource-exhausted cities in terms of innovation. On one hand, these policies significantly promoted innovation in resource-exhausted cities in the central and western regions, but their impact was less pronounced in eastern resource-exhausted cities. Specifically, cities that were reliant on oil showed significant policy effects, whereas those that were reliant on coal exhibited less clear effects. Therefore, the central government should comprehensively consider multiple factors in formulating targeted strategies for urban transformation. For instance, differential treatment in transfer payments should address specific challenges faced by resource-exhausted cities during their transformation, particularly by increasing support for those in the central and western regions. Additionally, the government should continuously assess the transformation performance of these cities through ongoing monitoring processes by evaluating aspects such as economic diversification, improvement in residents' living standards, and environmental remediation. Based on these assessments, establishing incentive mechanisms and gaining deeper insights into the strengths and weaknesses of each city type should serve as critical references for formulating differentiated policies. On the other hand, the magnitude of policy effects was also influenced by the talent environment and mining dependency. The lagging industrial base and lack of human capital in resource-exhausted cities constrained the innovation-promoting effects of the policies. This underscores the need for the government to address the potential mechanisms through which the talent environment and mining industry affect the innovation transformation of resource-exhausted cities. Future relevant policies should ensure the effectiveness of the support policy by improving industrial infrastructure and guaranteeing the supply of human resources.

Regarding the limitations of this study: First, the data used can be further refined. This study employed data at the prefecture level, but future research could use more granular county-level data to more comprehensively and meticulously evaluate the net effects of the support policy on urban innovation. Second, caution is needed when generalizing the conclusions of this study. China is the largest developing country in the world and has different economic and institutional environments compared with other developing countries. Therefore, when other countries seek to draw on China's empirical evidence to promote the transformation and development of their resource-exhausted cities, they must consider their own socio-economic contexts.

Supplementary Materials: The following supporting information can be downloaded from https://www.mdpi.com/article/10.3390/land13081153/s1: Table S1. Full list of pilot cities.

Author Contributions: Conceptualization, Y.T. and M.W.; methodology, D.Y.; software, D.Y.; validation, Z.H., Y.T., D.Y. and T.L.; formal analysis, Y.T and T.L.; investigation, Y.T.; resources, Y.T. and M.W.; data curation, D.Y.; writing—original draft preparation, Z.H., D.Y. and T.L.; writing—review and editing, Z.H., Y.T., D.Y. and M.W.; visualization, Z.H. and M.W.; supervision, M.W.; project administration, M.W. and Y.T.; funding acquisition, Y.T. All authors have read and agreed to the published version of the manuscript.

Funding: This research was supported by the 2022 National Social Science Fund Western Project entitled "Research on the Formation Mechanism and Improvement Strategy of Migrants Health Literacy" (grant number: 22XRK003).

Data Availability Statement: The data presented in this study are available on request from the corresponding author due to privacy restrictions.

Conflicts of Interest: The authors declare no conflicts of interest.

Notes

1 For detailed content, refer to the following, website: https://www.gov.cn/zwgk/2007-12/24/content_841978.htm (accessed on 5 June 2024).

2 For detailed content, refer to the following website: https://chinanews.com.cn/sh/2021/09-14/9564788.shtml (accessed on 10 July 2024).

3 For detailed content, refer to the following website: https://zfxxgk.ndrc.gov.cn/web/iteminfo.jsp?id=17391; https://www.gov.cn/zwgk/2007-12/24/content_841978.htm; https://www.gov.cn/zwgk/2013-12/03/content_2540070.htm (accessed on 10 July 2024).

References

1. Palmer, L. Green Structural Transformation. *Nat. Sustain.* **2023**, *6*, 877–878. [CrossRef]
2. Nelson, B.; Behar, A. *Natural Resources, Growth and Spatially-Based Development: A View of the Literature*; World Bank: Washington, DC, USA, 2012.
3. Weterings, R.; Bastein, T.; Tukker, A.; Rademaker, M.; de Ridder, M. *Resources for Our Future: Key Issues and Best Practices in Resource Efficiency*; The Hague Centre for Strategic Studies: The Hague, The Netherlands, 2013; ISBN 978-90-8964-529-6.
4. Sun, Y.; Liao, W.-C. Resource-Exhausted City Transition to Continue Industrial Development. *China Econ. Rev.* **2021**, *67*, 101623. [CrossRef]
5. Sachs, J.D.; Warner, A.M. The Curse of Natural Resources. *Eur. Econ. Rev.* **2001**, *45*, 827–838. [CrossRef]
6. Martinez-Fernandez, C.; Audirac, I.; Fol, S.; Cunningham-Sabot, E. Shrinking Cities: Urban Challenges of Globalization. *Int. J. Urban Reg. Res.* **2012**, *36*, 213–225. [CrossRef] [PubMed]
7. De Haas, R.; Poelhekke, S. Mining Matters: Natural Resource Extraction and Firm-Level Constraints. *J. Int. Econ.* **2019**, *117*, 109–124. [CrossRef]
8. Auty, R.M. Industrial Policy Reform in Six Large Newly Industrializing Countries: The Resource Curse Thesis. *World Dev.* **1994**, *22*, 11–26. [CrossRef]

9. Feng, Y.; Gao, Y.; Xia, X.; Shi, K.; Zhang, C.; Yang, L.; Yang, L.; Cifuentes-Faura, J. Identifying the Path Choice of Digital Economy to Crack the "resource Curse" in China from the Perspective of Configuration. *Resour. Policy* **2024**, *91*, 104912. [CrossRef]
10. Pack, H. Endogenous Growth Theory: Intellectual Appeal and Empirical Shortcomings. *J. Econ. Perspect.* **1994**, *8*, 55–72. [CrossRef]
11. Johnson, B. Cities, Systems of Innovation and Economic Development. *Innovation* **2008**, *10*, 146–155. [CrossRef]
12. Gutzmer, A. *Urban Innovation Networks*; Springer International Publishing: Cham, Switzerland, 2016; ISBN 978-3-319-24622-2.
13. Kalkanci, B.; Rahmani, M.; Toktay, L.B. The Role of Inclusive Innovation in Promoting Social Sustainability. *Prod. Oper. Manag.* **2019**, *28*, 2960–2982. [CrossRef]
14. von Schönfeld, K.C.; Ferreira, A. Urban Planning and European Innovation Policy: Achieving Sustainability, Social Inclusion, and Economic Growth? *Sustainability* **2021**, *13*, 1137. [CrossRef]
15. Zhang, J.-X.; Cheng, J.-W.; Philbin, S.P.; Ballesteros-Perez, P.; Skitmore, M.; Wang, G. Influencing Factors of Urban Innovation and Development: A Grounded Theory Analysis. *Environ. Dev. Sustain.* **2023**, *25*, 2079–2104. [CrossRef] [PubMed]
16. Wang, F.; Wong, W.-K.; Wang, Z.; Albasher, G.; Alsultan, N.; Fatemah, A. Emerging Pathways to Sustainable Economic Development: An Interdisciplinary Exploration of Resource Efficiency, Technological Innovation, and Ecosystem Resilience in Resource-Rich Regions. *Resour. Policy* **2023**, *85*, 103747. [CrossRef]
17. Namazi, M.; Mohammadi, E. Natural Resource Dependence and Economic Growth: A TOPSIS/DEA Analysis of Innovation Efficiency. *Resour. Policy* **2018**, *59*, 544–552. [CrossRef]
18. Ge, X.J.; Liu, X. Urban Land Use Efficiency under Resource-Based Economic Transformation—A Case Study of Shanxi Province. *Land* **2021**, *10*, 850. [CrossRef]
19. Luo, L.; Liu, P.; Zhu, F.; Sun, Y.; Liu, L. Policy Objective Bias and Institutional Quality Improvement: Sustainable Development of Resource-Based Cities. *Resour. Policy* **2022**, *78*, 102932. [CrossRef]
20. Baldwin, R.; Ito, T. The Smile Curve: Evolving Sources of Value Added in Manufacturing. *Can. J. Econ./Rev. Can. D'économique* **2021**, *54*, 1842–1880. [CrossRef]
21. Mudambi, R. Location, Control and Innovation in Knowledge-Intensive Industries. *J. Econ. Geogr.* **2008**, *8*, 699–725. [CrossRef]
22. Meng, B.; Ye, M.; Wei, S.-J. Measuring Smile Curves in Global Value Chains. *Oxf. Bull. Econ. Stat.* **2016**, *82*, 988–1016. [CrossRef]
23. Naudé, W.; Bosker, M.; Matthee, M. Export Specialisation and Local Economic Growth. *World Econ.* **2010**, *33*, 552–572. [CrossRef]
24. Capello, R.; Dellisanti, R. Smile-and-Go. Regional Performance through Global Value Chains in Europe. *Pap. Reg. Sci.* **2024**, *103*, 100018. [CrossRef]
25. Humphrey, J.; Schmitz, H. How Does Insertion in Global Value Chains Affect Upgrading in Industrial Clusters? *Reg. Stud.* **2002**, *36*, 1017–1027. [CrossRef]
26. Ang, Y.Y. *How China Escaped the Poverty Trap*; Cornell University Press: Ithaca, NY, USA, 2016.
27. Wang, S.; Yang, D.Y. *Policy Experimentation in China: The Political Economy of Policy Learning*; Policy Commons; National Bureau of Economic Research: Cambridge, MA, USA, 2021.
28. He, S.; Lee, J.; Zhou, T.; Wu, D. Shrinking Cities and Resource-Based Economy: The Economic Restructuring in China's Mining Cities. *Cities* **2017**, *60*, 75–83. [CrossRef]
29. Cai, F.; Wang, D.; Du, Y. Regional Disparity and Economic Growth in China: The Impact of Labor Market Distortions. *China Econ. Rev.* **2002**, *13*, 197–212. [CrossRef]
30. Su, K.; Li, J.; Fu, H. Smart City and the Applications. In Proceedings of the 2011 International Conference on Electronics, Communications and Control (ICECC), Ningbo, China, 9–11 September 2011; p. 1031, ISBN 978-1-4577-0320-1.
31. Zheng, S.; Kahn, M.E. China's Bullet Trains Facilitate Market Integration and Mitigate the Cost of Megacity Growth. *Proc. Natl. Acad. Sci. USA* **2013**, *110*, E1248–E1253. [CrossRef] [PubMed]
32. Chi, F.; Han, H. The Impact of High-Speed Rail on Economic Development: A County-Level Analysis. *Land* **2023**, *12*, 874. [CrossRef]
33. Lele, U.; Goswami, S. Agricultural Policy Reforms: Roles of Markets and States in China and India. *Glob. Food Secur.* **2020**, *26*, 100371. [CrossRef] [PubMed]
34. Li, X.; Zhuang, X. Eco-City Problems: Industry–City–Ecology, Urbanization Development Assessment in Resource-Exhausted Cities. *Sustainability* **2023**, *15*, 166. [CrossRef]
35. Erdoğan, S.; Çakar, N.D.; Ulucak, R.; Danish; Kassouri, Y. The Role of Natural Resources Abundance and Dependence in Achieving Environmental Sustainability: Evidence from Resource-Based Economies. *Sustain. Dev.* **2021**, *29*, 143–154. [CrossRef]
36. Voß, J.-P.; Smith, A.; Grin, J. Designing Long-Term Policy: Rethinking Transition Management. *Policy Sci.* **2009**, *42*, 275–302. [CrossRef]
37. Liu, X.; Li, Y.; Chen, X.; Liu, J. Evaluation of Low Carbon City Pilot Policy Effect on Carbon Abatement in China: An Empirical Evidence Based on Time-Varying DID Model. *Cities* **2022**, *123*, 103582. [CrossRef]
38. Guo, B.; Feng, Y.; Hu, F. Have Carbon Emission Trading Pilot Policy Improved Urban Innovation Capacity? Evidence from a Quasi-Natural Experiment in China. *Environ. Sci. Pollut. Res.* **2024**, *31*, 10119–10132. [CrossRef] [PubMed]
39. Xiao, H.; Zhang, L.; Ren, J. Sustainability Risks of Resource-Exhausted Cities in China: A Principal Component Analysis. In *Multi-Criteria Decision Analysis for Risk Assessment and Management*; Ren, J., Ed.; Springer International Publishing: Cham, Switzerland, 2021; pp. 225–249. ISBN 978-3-030-78152-1.

40. Lu, H.; Liu, M.; Song, W. Place-Based Policies, Government Intervention, and Regional Innovation: Evidence from China's Resource-Exhausted City Program. *Resour. Policy* **2022**, *75*, 102438. [CrossRef]
41. Zhang, Q.; Brouwer, R. Is China Affected by the Resource Curse? A Critical Review of the Chinese Literature. *J. Policy Model.* **2020**, *42*, 133–152. [CrossRef]
42. Cheng, J.; Zhao, J.; Zhu, D.; Jiang, X.; Zhang, H.; Zhang, Y. Land Marketization and Urban Innovation Capability: Evidence from China. *Habitat Int.* **2022**, *122*, 102540. [CrossRef]
43. Yang, D.; Wang, G.; Lu, M. Marketization Level, Government Intervention and Firm M&As: Evidence from the Local SOEs in China. *Appl. Econ. Lett.* **2020**, *27*, 378–382. [CrossRef]
44. Callon, M. Markets, Marketization and Innovation. In *The Elgar Companion to Innovation and Knowledge Creation*; Edward Elgar Publishing: Cheltenham, UK, 2017; pp. 589–609. ISBN 978-1-78254-852-2.
45. Newton, P.W. *Transitions: Pathways towards Sustainable Urban Development in Australia*; Springer Science & Business Media: Berlin/Heidelberg, Germany, 2008; ISBN 978-1-4020-8812-4.
46. Zhu, Y.; Luo, Y.; Chen, J.; Wan, Q. Industrial Transformation Efficiency and Sustainable Development of Resource-Exhausted Cities: A Case Study of Daye City, Hubei Province, China. *Environ. Dev. Sustain.* **2023**, 1–15. [CrossRef]
47. Antonelli, C. Innovation and Structural Change. *Économie Appliquée* **2002**, *55*, 85–120. [CrossRef]
48. Song, M.; Wang, S.; Fisher, R. Transportation, Iceberg Costs and the Adjustment of Industrial Structure in China. *Transp. Res. Part D Transp. Environ.* **2014**, *32*, 278–286. [CrossRef]
49. Wu, F.; Liu, R. Industrial Upgrading and Independent Innovation Ability Construction—Empirical Research Based on China's Provincial Panel Data. *China Ind. Econ.* **2013**, *5*, 57–69.
50. Wu, N.; Liu, Z. Higher Education Development, Technological Innovation and Industrial Structure Upgrade. *Technol. Forecast. Soc. Chang.* **2021**, *162*, 120400. [CrossRef]
51. Yin, K.; Miao, Y.; Huang, C. Environmental Regulation, Technological Innovation, and Industrial Structure Upgrading. *Energy Environ.* **2024**, *35*, 207–227. [CrossRef]
52. Utterback, J.M.; Suárez, F.F. Innovation, Competition, and Industry Structure. *Res. Policy* **1993**, *22*, 1–21. [CrossRef]
53. Acs, Z.J.; Audretsch, D.B. Innovation, Market Structure, and Firm Size. *Rev. Econ. Stat.* **1987**, *69*, 567–574. [CrossRef]
54. Yuan, R.; Wen, W. Managerial Foreign Experience and Corporate Innovation. *J. Corp. Financ.* **2018**, *48*, 752–770. [CrossRef]
55. Zheng, G.; Wang, S.; Xu, Y. Monetary Stimulation, Bank Relationship and Innovation: Evidence from China. *J. Bank. Financ.* **2018**, *89*, 237–248. [CrossRef]
56. Aghion, P.; Bloom, N.; Blundell, R.; Griffith, R.; Howitt, P. Competition and Innovation: An Inverted-U Relationship*. *Q. J. Econ.* **2005**, *120*, 701–728. [CrossRef]
57. Caragliu, A.; Del Bo, C.F. Smart Innovative Cities: The Impact of Smart City Policies on Urban Innovation. *Technol. Forecast. Soc. Chang.* **2019**, *142*, 373–383. [CrossRef]
58. Brandt, L.; Holz, C.A. Spatial Price Differences in China: Estimates and Implications. *Econ. Dev. Cult. Chang.* **2006**, *55*, 43–86. [CrossRef]
59. Brandt, L.; Van Biesebroeck, J.; Zhang, Y. Creative Accounting or Creative Destruction? Firm-Level Productivity Growth in Chinese Manufacturing. *J. Dev. Econ.* **2012**, *97*, 339–351. [CrossRef]
60. Beck, T.; Levine, R.; Levkov, A. Big Bad Banks? The Winners and Losers from Bank Deregulation in the United States. *J. Financ.* **2010**, *65*, 1637–1667. [CrossRef]
61. Liu, Q.; Qiu, L.D. Intermediate Input Imports and Innovations: Evidence from Chinese Firms' Patent Filings. *J. Int. Econ.* **2016**, *103*, 166–183. [CrossRef]
62. Bertrand, M.; Duflo, E.; Mullainathan, S. How Much Should We Trust Differences-In-Differences Estimates?*. *Q. J. Econ.* **2004**, *119*, 249–275. [CrossRef]
63. Petersen, M.A. Estimating Standard Errors in Finance Panel Data Sets: Comparing Approaches. *Rev. Financ. Stud.* **2009**, *22*, 435–480. [CrossRef]
64. Jiang, L.; Lu, Y.; Chen, Y. Is the Market Mechanism Conducive to the Disposal of Zombie Enterprises: A Case Study of the Relaxation of Foreign Capital Control. *World Econ.* **2018**, *41*, 121–145. (In Chinese)
65. Sun, T.; Lu, Y.; Cheng, L. Implementation Effect, Long-Term Mechanism, and Industrial Upgrading of Support policy for Resource-Exhausted Cities. *China Ind. Econ.* **2020**, *7*, 98–116. (In Chinese) [CrossRef]
66. Li, P.; Lu, Y.; Wang, J. Does Flattening Government Improve Economic Performance? Evidence from China. *J. Dev. Econ.* **2016**, *123*, 18–37. [CrossRef]
67. Kou, Z.; Liu, X. FIND Report on City and Industrial Innovation in China (2017), Fudan Institute of Industrial Development, School of Economics, Fudan University. 2017. Available online: https://ride.fudan.edu.cn/info/1020/1274.htm (accessed on 15 July 2024).
68. Liu, X.; Gao, T.; Wang, X. *Regional Innovation Index of China: 2017: How Frontier Regions Innovate*; Springer: Singapore, 2018; ISBN 9789811312045.
69. Zhu, C.; Lee, C.-C. The Effects of Low-Carbon Pilot Policy on Technological Innovation: Evidence from Prefecture-Level Data in China. *Technol. Forecast. Soc. Chang.* **2022**, *183*, 121955. [CrossRef]
70. Lu, Y.; Tao, Z.; Zhu, L. Identifying FDI Spillovers. *J. Int. Econ.* **2017**, *107*, 75–90. [CrossRef]

71. Abadie, A.; Diamond, A.; Hainmueller, J. Synthetic Control Methods for Comparative Case Studies: Estimating the Effect of California's Tobacco Control Program. *J. Am. Stat. Assoc.* **2010**, *105*, 493–505. [CrossRef]
72. Chen, Y.J.; Li, P.; Lu, Y. Career Concerns and Multitasking Local Bureaucrats: Evidence of a Target-Based Performance Evaluation System in China. *J. Dev. Econ.* **2018**, *133*, 84–101. [CrossRef]
73. Baron, R.M.; Kenny, D.A. The Moderator-Mediator Variable Distinction in Social Psychological Research: Conceptual, Strategic, and Statistical Considerations. *J. Pers Soc. Psychol.* **1986**, *51*, 1173–1182. [CrossRef] [PubMed]
74. Angrist, J.D.; Pischke, J.-S. *Mostly Harmless Econometrics: An Empiricist's Companion*; Princeton University Press: Princeton, NJ, USA, 2009; ISBN 978-0-691-12034-8.
75. Fan, G.; Wang, X.; Zhang, L.; Zhu, H. Report on the Relative Process of Marketization in Various Regions of China. *Econ. Res.* **2003**, *3*, 9–18+89. (In Chinese)
76. Yuan, B.; Li, C.; Xiong, X. Innovation and Environmental Total Factor Productivity in China: The Moderating Roles of Economic Policy Uncertainty and Marketization Process. *Environ. Sci. Pollut. Res.* **2021**, *28*, 9558–9581. [CrossRef] [PubMed]
77. Qiu, Y.; Wang, H.; Wu, J. Impact of Industrial Structure Upgrading on Green Innovation: Evidence from Chinese Cities. *Environ. Sci. Pollut. Res.* **2023**, *30*, 3887–3900. [CrossRef] [PubMed]
78. Démurger, S. Infrastructure Development and Economic Growth: An Explanation for Regional Disparities in China? *J. Comp. Econ.* **2001**, *29*, 95–117. [CrossRef]
79. Chen, M.; Zheng, Y. China's Regional Disparity and Its Policy Responses. *China World Econ.* **2008**, *16*, 16–32. [CrossRef]
80. Zhang, W.; Wang, D.; Yu, J. Research on the Development Path and Model of Alternative Industries in Resource-Based Cities. *Bull. Chin. Acad. Sci.* **2011**, *26*, 134–141. (In Chinese)
81. Azapagic, A. Developing a Framework for Sustainable Development Indicators for the Mining and Minerals Industry. *J. Clean. Prod.* **2004**, *12*, 639–662. [CrossRef]
82. Sun, L.; Abraham, S. Estimating Dynamic Treatment Effects in Event Studies with Heterogeneous Treatment Effects. *J. Econom.* **2021**, *225*, 175–199. [CrossRef]
83. Cai, H.; Liu, Q. Competition and Corporate Tax Avoidance: Evidence from Chinese Industrial Firms. *Econ. J.* **2009**, *119*, 764–795. [CrossRef]
84. Saleh, H.; Surya, B.; Annisa Ahmad, D.N.; Manda, D. The Role of Natural and Human Resources on Economic Growth and Regional Development: With Discussion of Open Innovation Dynamics. *J. Open Innov. Technol. Mark. Complex.* **2020**, *6*, 103. [CrossRef]
85. Iqbal, S.; Rasheed, M.; Khan, H.; Siddiqi, A. Human Resource Practices and Organizational Innovation Capability: Role of Knowledge Management. *VINE J. Inf. Knowl. Manag. Syst.* **2020**, *51*, 732–748. [CrossRef]

land

MDPI

Article

The Impact of High-Speed Rail on Economic Development: A County-Level Analysis

Fangting Chi [1] and Haoying Han [1,2,*]

1 College of Civil Engineering and Architecture, Zhejiang University, Hangzhou 310058, China; 22012149@zju.edu.cn
2 Center for Balance Architecture, Zhejiang University, Hangzhou 310058, China
* Correspondence: hanhaoying@zju.edu.cn; Tel.: +86-136-0651-1156

Abstract: High-speed rail has an important impact on the location choices of enterprises and the labor force, which is reflected in a complex space–time process. Previous studies have been unable to show the change characteristics between enterprises and the labor force at the county level. Therefore, based on the new economic geography theory, we first constructed a theoretical analysis framework to explore high-speed railway's impact on county economy development and then obtained the two economic subdivision factors' impacts: industrial enterprises and secondary labor force. Then, based on the panel data of 1791 county units in China from 2003 to 2019, the study constructed a multi-period PSM-DID model to empirically explore high-speed rail's impact on the county's agglomeration of industrial enterprises and secondary labor force. The results show that high-speed rail has a long-term negative effect on the county area's agglomeration of industrial enterprises. From the perspective of the labor force, high-speed rail has a long-term and continuous positive effect on the agglomeration of the secondary labor force in county units.

Keywords: high-speed rail; county territory; PSM-DID; industry; secondary labor force

1. Introduction

Transportation system infrastructure has historically been seen as a crucial component of regional economic development and as having a significant positive externality [1,2]. High-speed rail has grown in importance because of transportation infrastructure expansion [3]. By reducing the physical and temporal distance between cities, increasing the population, capital, information, and technology mobility, and improving the relative accessibility between cities, high-speed rail boosts the location benefits of stations and cities along the line [4,5]. According to studies, high-speed rail significantly impacts the local economy growth in terms of both industry and population [6–9]. Therefore, for regional economic integration, sustainable development, and reducing global poverty, qualitative and quantitative evaluation of the corresponding relationship between high-speed rail networks and population and industry is of great practical significance. The research on China's high-speed rail network's impact is important for planning and building the world's rail networks because China is a leading country in this field [10].

Scholars have domestically and internationally started to investigate high-speed rail networks' effects on regional economic development considering multiple factors such as land [11], immigration [12], employment [13], highly skilled labor [14], manufacturing enterprises[7], and service industry [15]. Research on high-speed train networks' regional effects in foreign literature dates back to 1967 [16]. Sands studied the regions in Japan where Shinkansen was introduced, and he concluded that high-speed rail encouraged population growth and flow [17]. Okamoto and Sato examined the Kyushu Shinkansen and concluded that opening high-speed rail lines increased land prices in metropolises to the detriment of smaller cities. According to Blum et al., regional corridor development, location restructuring of businesses and families, and economic functional zone specialization in Western industrialized countries were all facilitated by the high-speed rail network [18]. Chen and Hall investigated how the British InterCity 125/225 affected British economic geography, and one-, two-, and above-two-hour-away metropolitan regions affected by high-speed rail were considered the three key regional layers [19]. Heuermann and Schmieder investigated how the expansion of Germany's high-speed rail network affected workers' commute choices and concluded that it led people who lived in large cities to move to smaller ones [8]. In 2011, Willigers and Van Wee et al. conducted a representative study on the location choices of corporate offices in the Netherlands and proposed that high-speed rail would affect corporate office location choice and that high-speed rail accessibility would have a significant influence on the location choice of businesses, particularly knowledge-intensive businesses [20]. According to Diao M.'s analysis of China's "four vertical and four horizontal" high-speed rail networks, businesses can relocate from megacities to second-tier cities near high-speed rail corridors thanks to intercity trade, labor mobility, and knowledge spillover [21]. The relationship between a high-speed rail network and local economic growth has also been examined by both domestic and international experts at many scales, including the European [3], national [12], urban agglomeration [6], provincial, and city levels. In conclusion, studies on high-speed rail networks' economic impacts typically focus on examining a single element, and the research scale neglects providing small- and medium-sized towns and counties the attention they require.

High-speed rail's "siphon effect" and "trickle effect" counterbalance the economic growth of urban units. The "siphon effect" describes how, following the opening of high-speed rail, the great allure of big cities is more likely to draw resources, labor force, and businesses from smaller cities along the route, thus harming the growth of small- and medium-sized cities. The term "trickle effect" refers to how larger surrounding cities aid smaller ones through consumption, employment, industry, and other factors to close the regional imbalance. Due to its impact on the supply and turnover of the professional labor market, high-speed rail has grown to be the most complicated issue in firms' location choices [9]. Most studies on how rail affects labor have concentrated on rural-to-urban migration, neglecting the "trickle effect" of high-speed rail's exodus of skilled people to neighboring counties. The high-speed rail region will increase workforce mobility and draw more businesses that require large and specialized labor forces. Obviously, the "siphon effect" has significantly and positively impacted the growth of key cities' economies. However, we cannot determine the exact effects of the "siphon effect" and "trickle effect" on economic growth and population mobility for other urban units (county-level cities and counties) that have high-speed rail based on impressions.

Here, we carry out an empirical study on 1791 county units in China and provide a theoretical framework based on new economic geography theory. China is unquestionably an innovator in high-speed trains. China's high-speed rail system, with a total length of 19,000 km, surpassed the rest of the world's network in length by 2015. China's high-speed train network continues to reach thousands of counties regardless of the country's vastness. The study on the effects of high-speed rail is typical in that it covers a large sample and a long period of time and has reference value for the transportation and economic development of small- and medium-sized cities both domestically and overseas, as well as for new urbanization and suburbanization.

This study's aims were to: (1) Investigate the interactive mechanism of the county's population and business being impacted by the high-speed rail network. (2) Analyze high-speed rail's influence on the county's secondary industries and the labor force agglomeration of industrial enterprises. (3) Answer the question, what impact does the introduction of high-speed rail have on the county's economic growth mode and development transformation?

2. Analysis Framework

The inverted "U"-shaped curve in the core–edge model proposed by Krugman proves that, under the interaction of increasing return to scale, population flow, and transportation cost, the forward and backward industry correlation effect is the strongest when the transportation cost is at the intermediate level [22]. The path dependence and iceberg transport cost proposed by the new economic geography theory demonstrate transportation's importance in the industrial agglomeration process. The high-speed rail network's economic impact encourages the movement and concentration of companies and staff. The high-speed rail network's impact on the county economies' evolution has received a lot of attention from the academic community because of the extensive high-speed rail network building in China's counties. The relationship between "high-speed rail and population", "high-speed rail and industry", "high-speed rail and relative accessibility", and other discrete elements or with urban spatial structure has been examined and analyzed in previous research. However, a particular regional economy's growth is the consequence of the coordinated actions of numerous complex elements, and a high-speed rail network's impact on a county economy's growth cannot be presented scientifically and precisely through a cursory investigation of a single aspect. Here, we look at the correlation between the high-speed rail network and county economic growth. Starting with the two economic components of businesses and the labor force, we investigate the mechanism of high-speed rail networks on county economic development based on new economic geography theory.

2.1. The Relationship between High-Speed Rail Network and County Economic Growth

County units are all part of a complex urban network system from the perspective of the overall regional spatial structure. The connections with other urban units and the comparative advantages with units of the same rank are important decisive factors for economic development. The local advantages, industrial characteristics, and economic foundation of county units will affect their subsequent economic development's quality and speed. Introducing high-speed rail widens the pathway for factor circulation, and the interaction demand for factor flow among urban units encourages the creation and development of high-speed rail which, in turn, supports and directs urban unit development and impacts the distribution trend of enterprises and labor factors. We developed a flow model of four types of urban units (high-speed urban area, non-high-speed urban area, high-speed county, and non-high-speed county) and two types of economic elements (enterprises and labor force) under the influence of high-speed rail based on this and in combination with pertinent examples in the literature (Figure 1). The focus is on investigating high-speed rail's effects on the flow of economic components of county units and on a more intuitive study of the interplay between the two elements and urban unit objects.

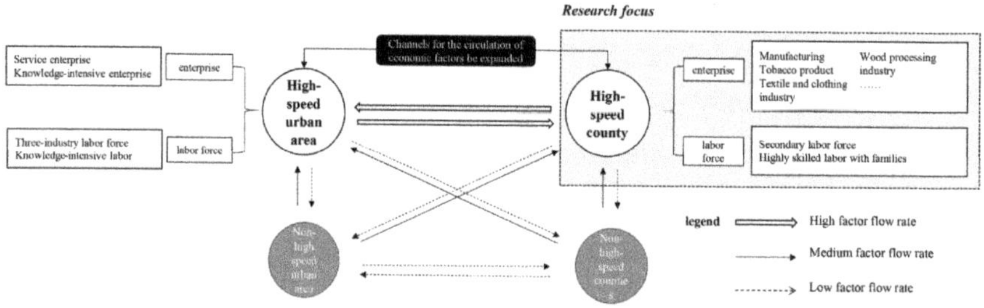

Figure 1. Schematic diagram of flow model of enterprise and labor force factors under high-speed rail's influence.

The relevant statistics and literature reviews demonstrate that high-speed rail broadens the production factor's circulation channel and brings it spatially and temporally closer to the center metropolis. According to the analysis of the factor flow model in Figure 1, two aspects are mostly responsible for luring businesses to the high-speed railway county. On the one hand, enterprises are forced to relocate to counties from the urban areas that high-speed rail has opened up due to the knowledge spillover effect of central cities [14] and congestion costs [7,21], particularly low-end manufacturing, textile, and other industrial enterprises. Because central cities are increasingly clustered in terms of population, information and capital do not directly participate in production and modern service and knowledge-intensive sectors tend to be concentrated there [15,23]. On the other hand, the impetus comes from urban units without high-speed rail service, as the expansion of the consumer market [13], the specialized labor market [24,25], and the product input channels [26] encourage businesses to group together in high-speed rail counties. Similarly, the driving force to attract labor force in a high-speed rail county mainly comes from two aspects: on the one hand, the secondary labor force seeks employment possibilities, lowers living expenses, and considers family emigration [8,12]. Meanwhile, high-speed trains also increase the externality of human capital, and major cities are where talents, innovations, and ideas prefer to congregate [27]. On the other hand, the second and third industrial workforces from non-high-speed rail towns relocate there for various reasons, including job searching and lifestyle improvement [28].

Here, we look at the agglomeration tendency that high-speed rail has had on the two economic components of businesses and labor. Additionally, the two economic forces that are most likely to congregate in counties served by high-speed rail can be separated into industrial firms and the accompanying secondary labor force, according to the flow model study. The following research's key emphasis is the relationship between the two economic forces and high-speed rail opening.

2.2. Effect Mechanism of High-Speed Rail Network on County Economic Development

Exploring the mechanism between the high-speed rail network and corresponding economic growth is of long-term significance to county economies' development and transformation. As a key conduit for economic factor movement, such as inter-regional capital, information, and human flow, the high-speed rail network reconstructs the market share of factor flow. We examine the dynamic mechanism underlying the county economy's development and transformation path under the influence of high-speed railway, focusing on the two main economic segmentation factors of industrial enterprises and secondary labor force based on the findings in the previous section (Figure 2).

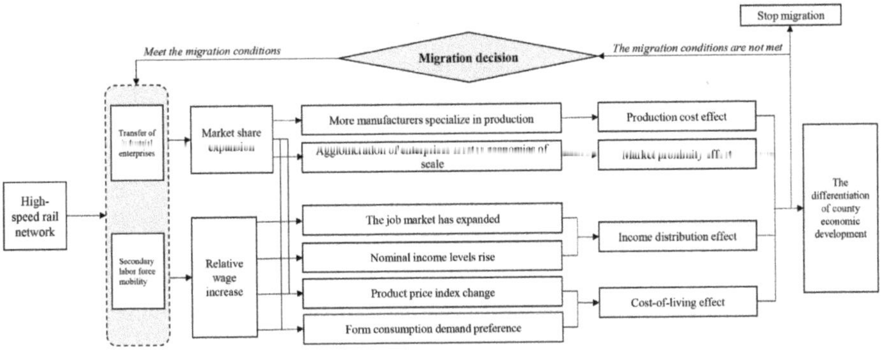

Figure 2. Action mechanism diagram of high-speed rail network on county economic development and transformation path.

For industrial enterprises, counties with high-speed rail may draw industrial business clusters. The market share increase enables more manufacturers to carry out more specialized production, and the grouping of businesses creates economies of scale, thus reducing production costs [29] and improving market access [30], both of which encourage industrial industry growth in counties. Regarding the secondary labor force, the high-speed rail network speeds up labor factor movement, the decline in living expenses such as housing costs, rent, consumption, and education increases relative wages, and the concentration of industrial enterprises also widens the job market and impacts income distribution [14]. The cost of living is lowered and the cost of living effect is produced by changes in the product price index and customer demand preferences [31].

In conclusion, industrial businesses and secondary labor force distribution restructuring and agglomeration will collaborate to foster regional economy development independent of geographic and spatial considerations. Nevertheless, because of the unique characteristics of high-speed rail counties in the urban system, the "siphon effect" of central cities produced by high-speed rail may prevent industrial enterprises and the secondary labor force from congregating in high-speed rail counties, and it may even cause enterprises and populations that were originally located in high-speed rail counties to reverse to central cities. Thus, the economic variables most likely to congregate in counties opened by high-speed rail are industrial companies and the secondary labor force. The precise agglomeration trend, however, cannot be determined with sufficient accuracy based on theoretical study because of central cities' influences. Thus, high-speed rail counties must necessarily undergo reasonable economic development.

2.3. Exploration of Different Evolutionary Paths of High-Speed Railway Counties

Introducing high-speed rail considerably boosted the high-speed rail counties' accessibility to high-speed rail urban areas when compared to non-high-speed urban areas and counties. Regarding industrial enterprises, the cost effect of lower factor flow costs and the market proximity effect of a wider potential market scope encourage the establishment of new businesses. Consequently, the labor force demand first exceeds supply, and the county is transformed into a satellite city formed by industrial agglomeration, into an "auxiliary city". Regarding the labor force, the county will transform into a satellite city formed by population agglomeration, into a "sleeping city", because the secondary labor force supply leads the settlement of industrial enterprises due to the living cost effect caused by the increase in new service supply and the income distribution effect caused by the expansion of potential employment opportunities. The mechanism and differentiation of the impact of high-speed rail opening on high-speed rail counties' transformation and development are shown below (Figure 3).

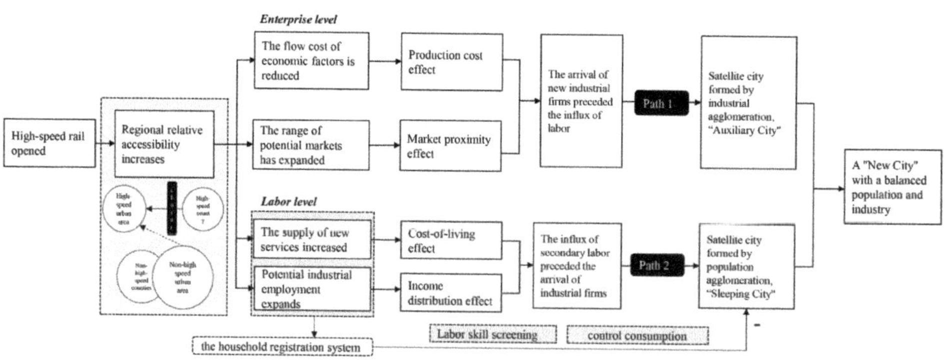

Figure 3. Mechanism and differentiation of high-speed railway opening affecting transformation and development of high-speed railway counties.

Several academics have also proposed that counties with high-speed rail systems attract new businesses and citizens. The core of industrial transfer is the process of business relocation and location adjustment. New businesses locate themselves in high-speed railway counties for various reasons. The primary factor is that secondary industries such as manufacturing and industry in developing nations typically originate in the major towns [32]. To save money on land and labor, businesses are relocating to nearby towns [33]. Additionally, knowledge-intensive businesses have high location and transportation needs [34] and a tendency to congregate in high-speed rail counties due to their high technology and demand for information exchange. The primary driver for people to relocate to high-speed rail counties is the significant decrease in immigration costs, which encourages the movement of labor from relatively underdeveloped areas to these counties and the significant reduction in commute times, which increases employment opportunities and decreases living expenses [35].

2.4. Summary of the Chapter

To further address the research questions posed in the introduction, the chapter explores the mechanism by which the high-speed rail network affects county economic development. It also examines the agglomeration and exodus of businesses and workers in four different types of urban units (high-speed urban area, non-high-speed urban area, high-speed county, and non-high-speed county). High-speed rail opening in a county widens the channels of factor circulation, thus bringing high-speed rail counties closer to the center urban region in terms of both time and geography, according to the analysis framework. The high-speed railway significantly impacts industrial business movement and concentration as well as the secondary labor force in the county area. The original equilibrium state is disrupted by the high-speed railway's opening, and businesses and the labor force decide to migrate accordingly. It creates a fresh chance for the high-speed railway county to change and develop. We conjecture there are two primary paths for high-speed railway county transformation and growth based on analyzing enterprise and labor force levels: the first route is the formation of a satellite city formed by industrial agglomeration, the "auxiliary city"; another path is formation of a satellite city of population agglomeration and development by the influx of a large number of non-agricultural laborers, the "sleeping city".

3. Research Design

3.1. Research Problem

According to the analysis framework, from the standpoint of industrial enterprises, a high-speed rail opening greatly lowers industrial companies' transportation costs in county regions, thus affecting market access and influencing production costs. However, it also heightens the center urban area's siphon effect, which increases industrial businesses' propensities to relocate to the central city with a higher total external income. Here, we propose hypothesis 1: under the current national circumstances, a high-speed rail opening will have long-term adverse effects on industrial companies at the county level and weaken industrial enterprises' agglomeration level at the county level. Regarding the secondary labor force, a high-speed rail opening lowers living expenses and transportation costs and creates a cost-of-living effect to draw laborers into the central metropolitan region. On the other hand, the availability of new infrastructure services and the expansion of employment prospects affect income distribution and support the transformation of the agricultural population into a non-agricultural one. Here, we propose hypothesis 2: introducing high-speed rail will have a long-lasting and positive effect on the county's secondary labor force concentration. Additionally, hypothesis 3 is proposed considering the conclusion of the counties' development and transformation from the perspective of high-speed rail opening in the analysis framework 2.3: a satellite city formed by population agglomeration, "sleeping city" is formed because of the high-speed rail opening, which alters counties' economic growth styles. Industrial enterprise settlement precedes the influx of the secondary labor force.

Based on the panel data of 1791 county-level urban units (county-level cities and counties) across the nation of China from 2003 to 2019, this chapter employs the multi-period PSM-DID model to conduct the appropriate research design to test the three assumptions.

3.2. Method Selection

The selection bias issue with county unit samples was successfully resolved using the propensity score matching technique (PSM). Using the differential difference, the "difference-in-differences" analysis (DID) effectively examined the policy impact of high-speed rail on the county unit while resolving the endogeneity issue engendered by the incomplete dependent variables of the county unit itself during the research process. The individual differences between the various locations prior to the high-speed rail inauguration will be disregarded by using the cross-section model and the time effect will be disregarded by using the time series model to avoid the constraints of the single difference technique. To evaluate high-speed rail's effect on county economic development, we primarily draw on the methodological work of Goodman-Bacon (2021) [36] and Callaway and Sant 'Anna (2022) [37] and uses of the multi-time point differential approach (DID, differences in differences). Here, the propensity score matching multiple difference approach (PSM-DID) was employed to more precisely and effectively assess the reconstruction of county economic space by high-speed rail. The specific operations and models are as follows:

1. To identify the group that corresponds with the county units that offer high-speed rail service, use PSM.
2. The county units of the matched experimental group and control group were used for the DID model evaluation and analysis, and the following regression equation was obtained after combination:

$$Y_{it}^{PSM} = \beta_0 + \beta_1 city_i + \delta_0 post_{it} + \delta_1 city_i * post_{it} + \beta_2 xlist_{it} + \mu_i + \lambda_t + \varepsilon_{it} \quad (1)$$

where i denotes the region, t represents the year, and Y_{it} is the dependent variable, indicating the economic development level of county i in the year t (including the agglomeration level measurement of the industrial enterprises and the secondary labor force's agglomeration degree). $city_i$ is an individual dummy variable of the county. If the policy of the high-speed rail opening affects county i, its value is 1, while the county not affected by the policy is 0. $post_{it}$ represents the time dummy variable in the processing period of the high-speed railway policy; the value of the county is 1 after the high-speed railway is opened and 0 if it is not. The interaction item ($city_i * post_{it}$) represents the dummy variable of the county unit after the high-speed railway opening. Its coefficient δ_1 is the difference between the impact of the high-speed railway opening on the treatment group and the control group, which this paper focuses on. $xlist_{it}$ is a group of control variables affecting the county unit economic development, and μ_i represents the individual fixed effect and is used to control the heterogeneity of the county units, while λ_t represents the time fixed effect and is used to control the corresponding year of county units, and ε_{it} is the residual phase.

3.3. Variable Selection

There are two types of indicators for county unit economic growth: multiple indicators and comprehensive single indicators. The study examines high-speed rail's effects on the urban unit economic development using the indicators used by Ahlfeldt and Feddersen [38], Redding and Tumer [39], and Kim [35]. It also confirms the above hypothesis regarding the patterns of enterprises and labor force agglomeration in county economic development. The multi-index analysis method was employed in assessment. We focused on how high-speed rail affects industrial firm dispersion as well as labor force movement and concentration in secondary sectors such as manufacturing. Several academics both domestically and internationally utilize the percentage of the total industrial output value to gauge the degree of regional industrial agglomeration when studying the production and agglomeration of county enterprises. For instance, to determine the industrial agglomeration level, Wen (2004) [40] and Yu Jin (2006) [41] both used the percentage of each region's industrial production value in the overall industrial GDP of the year as a variable. Since the data on the country's gross industrial output value were only available through 2012, the paper substituted the GDP for the corresponding year. Implementing the index was comparatively reasonable and dependable because the tangible and quantifiable aspects of manufacturing items have been perfected in statistical caliber and procedures. The proportion of workers in secondary industry units across the entire nation at the end of the year was used to gauge the industrial labor force agglomeration degree in terms of the size, distribution, and concentration of the county labor force.

The DID model was employed to address some of the study object's endogeneity issues. However, to more precisely analyze the alterations and variations of the industrial enterprise and secondary labor agglomeration levels across counties with high-speed and non-high-speed rail, we cite the study by Shao et al. [6] on service sector agglomeration in the Yangtze River Delta region of China, the study by Dai and Hatoko [42] on the economic disparities between Switzerland and Japan regarding high-speed rail, and the study by Wang et al. [43] on high-speed rail service's effects on population flow and urbanization regarding industry and population by using control variables to fix the model's endogenous and sequential issues. In addition, the control variables were treated logarithmically to eliminate the collinearity issue. The final selection of control variables included the county's total population at the end of the year (lnpop), the gross regional product (lngdp), the national market potential (lnpot), the local market potential (lnlopot), and the completed amount of urban fixed assets investment (lnfid). To govern county economic development, industrial enterprise agglomeration, and secondary industry labor agglomeration, these factors were used as variables. For a description of the chosen variables and their selection, see (Table 1).

Table 1. Variable description.

Classification	Variable	Symbol	Unit	Definition
Dependent variable	Agglomeration level of industrial enterprises	sec	%	Local industrial output value above designated size/annual GDP
	Concentration degree of secondary labor force	emp	%	Local employees in secondary industry units at the end of the year/national employees in secondary industry units at the end of the year
Independent variable	County unit dummy variable	$city_i$	/	By 2019, the value of county units with high-speed rail service will be 1, otherwise 0
	Policy processing period dummy variable	$post_{it}$	/	From 2003 to 2019, the county value was 1 after the high-speed rail service and 0 before the high-speed rail service
	Interaction item	$city_i * post_{it}$	/	After the high-speed rail line opens, the virtual variable in the county area is 1 in terms of time dimension; otherwise, it is 0. The virtual variable of the county area opened by high-speed rail is 1, else it is 0 in terms of region dimension
Control variable	Total population of the county at the end of the year	lnpop	Ten thousand people	County population at the end of the year, logarithm
	Gross regional domestic product	lngdp	Ten thousand CNY	Gross county product of corresponding year, logarithm
	National market potential	lnpot	/	The logarithm of the total retail sales of consumer goods in the local county divided by the sum of distances from other urban units (urban area, county area) to the local county unit ($mp_i = \sum_{j=1}^{R} RET_j d_{ij}^{-1}$, where mp_i is the market potential of county *i*, *RET* represents the total retail sales of social consumer goods in county, and d_{ij} is the distance between county *i* and county *j*
	Local market potential	lnlopot	/	The logarithm of the total retail sales of consumer goods in a county divided by the distance to the nearest prefecture-level city ($lomp_i = RET_i/Nd_{ij}$, where $lomp_i$ is the local market potential of county i, RET represents the total retail sales of consumer goods in county, and Nd_{ij} is the distance between county i and the nearest prefectural-level city j
	Total investment in urban fixed assets completed	lnfid	/	The amount of urban fixed assets investment completed in the corresponding year is logarithm

Notes: sec, lnpop, lngdp, and lnfid all use 10,000 as the unit of measurement in variable calculation; since the number of secondary industry employees in the county is relatively small, the emp variable uses each person as the unit of measurement; in lnpot and lnlopot, the total retail sales of social consumer goods are measured in 10,000, and the distance is measured in kilometers.

3.4. Description of Research Objects and Data

3.4.1. Research Object and Scope

The essay primarily investigates how high-speed rail's launch has affected local economic growth (focusing on the interaction between industrial production and the corresponding secondary labor force). Thousands of county-level administrative divisions in China are becoming increasingly crucial with the declining market (Source: National Development and Reform Commission, China, https://www.ndrc.gov.cn/ (accessed on 1 July 2022)). A city's formation and growth cannot progress without considering administrative power. In mainland China, the term "urban establishment" refers to the administrative establishment system, which includes cities, counties, and municipalities directly under the central government. As of March 2020, there were 333 prefecture-level administrative divisions in China (excluding Hong Kong, Macao, and Taiwan) (Source: http://data.acmr.com.cn/member/city/city_md.asp (accessed on 28 July 2022)). These 300-plus cities have received the most attention in studies on China's high-speed rail network's effects [44–49]. We consider that county-level administrative divisions have evolved into the initial transfer point for non-agricultural enterprises and population agglomeration, because of China's huge area, enormous population, and the high-speed rail network constantly spreading and encompassing counties. Researchers studying high-speed rail should note China's thousands of county-level administrative units, which are crucial to the country's development. Geographically, economically, and in terms of management authority, the municipal district is closer to the central city than the county-level cities and counties under its jurisdiction; we list the primary urban units examined in this article as the county-level cities and counties governed by the central cities. Based upon this, the nation's cities are split into four categories of urban units, including high-speed urban area, non-high-speed urban area, high-speed county, and non-high-speed county, according to the administrative zoning size. The four categories of urban units are explained in detail below, and the geographic spatial distribution of the four types of urban units during the research year (up to 2019) is visualized in Arcgis10.8 (Figure 4) (Standard map base source: http://bzdt.ch.mnr.gov.cn/ (accessed on 30 July 2022)). To investigate the economic growth trajectory and population distribution rule of high-speed rail counties under the influence of high-speed rail, we selected 1791 county-level urban units (county-level cities and counties) across the country as empirical research subjects. The four types of urban units were classified as follows:

1. High-speed urban area: If any of the sub-provincial-level cities and prefecture-level city districts opened high-speed railways, the whole prefecture-level city district was divided into high-speed railway central city units, such as Beijing, Shanghai, Tianjin, and Chongqing.
2. Non-high-speed railway urban areas: In contrast to the high-speed urban areas, the prefecture-level city municipal districts were divided into non-high-speed railway urban units if there was no high-speed service in any section of the sub-provincial city or prefecture-level city municipal district.
3. High-speed county: In addition to the central city units mentioned above, if county-level cities, county-level administrative units, and the units below had high-speed rail service, they were classified as high-speed rail counties.
4. Non-high-speed county: Conversely, if county-level cities, county-level administrative units, and the units below had no high-speed rail service, they were classified as non-high-speed counties.

3.4.2. Data Sources

The panel data of county area city units, constituted of 1791 counties and county-level cities under the jurisdiction of the sample urban areas in China from 2003 to 2019, were used as the county data in this article. The administrative region zoning by the end of 2020 was used as the benchmark, considering China's administrative divisions. Regarding high-speed rail, information about China's high-speed rail projects from 2003 to 2019 was gathered and organized using the websites for "Train Schedule 2010," "Train Schedule 2020," the National Railway Administration of China, and the High-Speed Railway Network. For details on data gathering, see the table below (Table 2).

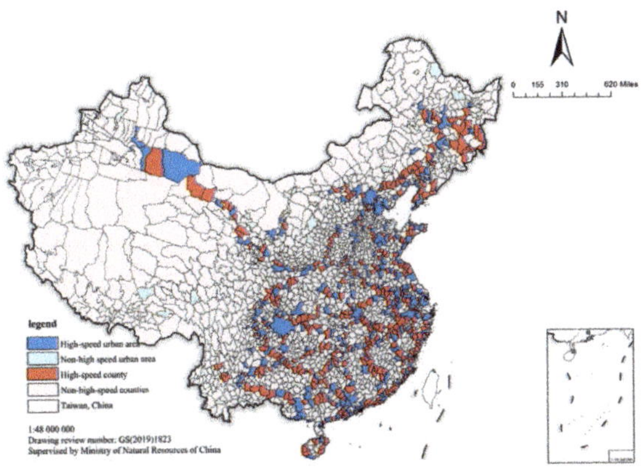

Figure 4. Geographical spatial distribution of four types of urban units (by the end of 2019). Note: This map is based on the standard map No. GS (2019)1823 downloaded from the Standard Map Service website of the Department of Natural Resources. The base map is unchanged.

Table 2. Summary table of datasets involved in the study.

Data Set	Data Sources	Year
China County Statistical Yearbook (County and City Volume)	Summary by Office for National Statistics: www.stats.gov.cn (accessed on 1 July 2022).	2003–2019
Base map of county-level administrative divisions in China	Chinese government website: www.gov.cn(accessed on 19 July 2022)	By 2020
County city unit base map	Based on county-level administrative division file arrangement	By 2020
City centers of prefecture-level cities and counties	Summary by Chinese government website: www.gov.cn (accessed on 17 July 2022).	By 2020
The distance between counties and urban units	Calculation based on Arcgis10.8 software	By 2020
Timetable of high-speed rail	Summary by China Railway Network: www.12306.cn (accessed on 17 July 2022).	2010, 2020
Data of the opening of high-speed rail lines over the years	Summary by National Railway Administration: www.nra.gov.cn; High-speed rail network: www.gaotie.cn (accessed on 14 July 2022)	2003–2019
Data of the opening of high-speed rail stations over the years	Summary by National Railway Administration: www.nra.gov.cn; High-speed rail network: www.gaotie.cn (accessed on 14 July 2022)	2003–2019

3.4.3. Sample Matching

We performed PSM matching between the treatment and control groups prior to the empirical analysis. Consequently, 358 county units that had opened high-speed rail between 2003 and 2019 constituted the treatment group, while 1433 county units that had not opened high-speed rail during that time constituted the control group. The probit model was used to estimate the P-score. The weight was calculated using the nearest-neighbor tendency matching approach, and the "on support" condition was appended (see Appendix A (Tables A1–A3 and Figure A1) for specific PSM results).

The kernel density function's distribution curve before and after the nearest-neighbor tendency was drawn (see Figure 5). According to Figure 5a in the comparison chart, the distribution of counties in the treatment group was extremely loose before PSM matching was carried out. In contrast, the control group's P-score kernel density distribution was clearly concentrated and skewed to the left. The impact of high-speed rail on the P-score probability density distribution in counties differed widely. The results of the multi-stage DID analysis would unavoidably have been seriously affected by sample selection bias if PSM matching were not used and the concentration level of the industrial enterprises and the secondary industry labor force between the two groups of the county samples were directly compared. The findings' reliability would thus be impacted. We compared the characteristics of the two county sample groups in five dimensions: GDP, market potential, distance from the county center to the closest prefecture-level city, and the amount of urban fixed assets investment that was completed at the end of the year. Furthermore, we employed the nearest-neighbor method. After matching, the retained samples' features were identical in every way, and the selectivity bias was essentially removed (see Figure 5b).

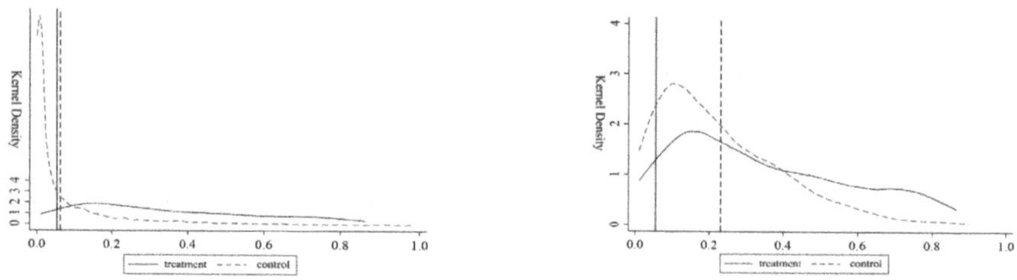

(a) (b)

Figure 5. Comparison of kernel density distribution of propensity scores between the treatment group and the control group before and after nearest-neighbor propensity matching. (**a**) Kernel density of propensity score before matching. (**b**) Kernel density of propensity score after matching).

4. Empirical Analysis

4.1. Impact Analysis of High-Speed Rail on County Economic Development Based on PSM-DID

The section focuses on the heterogeneity of high-speed railway construction's influence on the labor force agglomeration of industrial enterprises and secondary industries. To test the PSM sample matching's effectiveness, a mix of OLS baseline regression 1 (see model 1), OLS baseline regression after PSM matching (see model 2), fixed-effect model baseline regression (see model 3), DID regression using samples satisfying the common support, which is the tendency of scores of the treatment group and the control group to have a large common range of values (see model 4), multi-phase PSM-DID regression, and samples satisfying the common supporting hypothesis (see model 5) were used. After PSM-DID estimation, the standard error mean of the control variables was reduced, and the interaction term's regression coefficient was larger than that of the ordinary DID estimated by the fixed-effect model's regression. Overall, we think that PSM-DID's estimation effect was better.

4.1.1. Impact of High-Speed Rail on Industrial Agglomeration in Counties

The section focuses on the industrial agglomeration differences between counties served by and those not served by high-speed rail (see Table 3). The following outcomes were found: (1) With the exception of model 1, all five models' regression coefficients for the interaction terms of the explanatory variables (city*post) were significant above the 10% level, showing that the regression results were still reliable even after accounting for the selectivity bias issue. The interaction terms of the matching results of PSM-DID were also significant at the 1% level. This demonstrates how individual variances between counties can skew the research findings. (2) Qin's research (2017) also points out that high-speed railway construction reshapes counties' economic activities. Moreover, it has certain negative effects on county economic development [50]. (3) The regression results of the control variables show that the gross regional product, the size of the national market, and the completion of urban fixed asset investments have a significant impact on the industrial agglomeration in counties, whereas the total population at the end of the year and the size of the local market have little effect. The possible reason is that a county's industrial development is not only related to the nearest central city, but also closely related to its own level of economic development, urban construction, and market potential in a larger scope.

4.1.2. High-Speed Rail's Influence on the Agglomeration of Secondary Industry Labor Force in Counties

The section focuses on the differences in secondary labor agglomeration between counties with and without high-speed rail because of high-speed rail installation (see Table 4). The following are the outcomes: (1) The regression coefficient of model 5 shows obvious significance at the 5% level, indicating that high-speed rail opening has a significant impact on counties' secondary industry labor force agglomeration. Meanwhile, the regression coefficient is positive; indicating that, in contrast to the result of the industrial enterprises, high-speed railway opening promotes the concentration of secondary labor force in the county. The possible reason is that the county area attracts the secondary labor force of the central city and the surrounding agricultural population. (2) Among the control factors, the size of the fixed asset investment in metropolitan areas, the size of the national market potential, and the size of the local market potential all significantly affect the labor force distribution in the secondary industry. The secondary industrial labor force concentration in the county is also not significantly impacted by the total population at the end of the year or the gross regional product. In contrast, it is clear from the coefficient that the local market potential has a very beneficial effect on the secondary industry labor force concentration in the county when compared to the degree of industrial agglomeration.

Table 3. Evaluation results of impact of high-speed rail opening on industrial agglomeration in county area based on multi-phase PSM-DID model.

	Model 1	Model 2	Model 3	Model 4	Model 5
	ols1	ols2	fe	Common Support	Psm-Did
citypost	−0.00013	−0.00015 ***	−0.00005 *	−0.00005 *	−0.00007 ***
	(−1.52423)	(−4.12649)	(−1.74838)	(−1.74838)	(−3.21568)
lnpop	−0.00011 ***	−0.00007 *	−0.00022 ***	−0.00022 ***	−0.00008
	(−3.96557)	(−1.81789)	(−4.33313)	(−4.33313)	(−1.20055)
lngdp	0.00036 ***	0.00115 ***	0.00020 ***	0.00020 ***	0.00072 ***
	(6.20586)	(21.13444)	(6.38093)	(6.38093)	(9.97349)
lnpot	0.00090 ***	0.00010	−0.00026 ***	−0.00026 ***	−0.00049 ***
	(3.58495)	(1.07191)	(−3.14396)	(−3.14396)	(−1.0 × 10^2)
lnlopot	−0.00002	0.00002	−0.00003	−0.00003	0.00005
	(−0.89649)	(0.91598)	(−1.26638)	(−1.26638)	(0.61204)
lnfid	−0.00015 ***	−0.00054 ***	0.00000	0.00000	−0.00006 ***
	(−5.44759)	(−1.8 e + 01)	(0.04726)	(0.04726)	(−3.01223)
N	3706	833	3706	3706	833
Adj. R^2	0.50289	0.57250	0.27987	0.27987	0.37081

Note: Numbers in brackets are standard error. *** and * are significant at the level of 1, and 10%, respectively.

Table 4. Evaluation results of impact of high-speed rail opening on secondary industry labor force agglomeration in county based on multi-phase PSM-DID model.

	Model 1	Model 2	Model 3	Model 4	Model 5
	ols1	ols2	fe	Common Support	Psm-Did
citypost	−0.00006 ***	−0.00004 *	0.00002	0.00002	0.00004 **
	(−3.51918)	(−1.78358)	(1.01943)	(1.01943)	(2.28175)
lnpop	0.00010 ***	0.00021 ***	−0.00001	−0.00001	0.00008
	(12.04140)	(8.70324)	(−0.29664)	(−0.29664)	(1.59268)

Table 4. *Cont.*

	Model 1	Model 2	Model 3	Model 4	Model 5
	ols1	**ols2**	**fe**	**Common Support**	**Psm-Did**
lngdp	0.00016 *** (14.78234)	0.00063 *** (17.36344)	0.00004 ** (2.40126)	0.00004 ** (2.40126)	0.00008 (1.33910)
lnpot	0.00108 *** (41.40345)	0.00060 *** (9.34073)	−0.00005 (−0.69452)	−0.00005 (−0.69452)	−0.00023 *** (−6.28128)
lnlopot	−0.00001 ** (−2.26344)	−0.00002 (−1.13555)	0.00002 (1.02721)	0.00002 (1.02721)	0.00012 ** (2.00929)
lnfid	−0.00010 *** (−1.8 × 10^2)	−0.00036 *** (−1.8× 10^2)	0.00001* (1.90752)	0.00001 * (1.90752)	−0.00003 * (−1.82140)
N	3706	833	3706	3706	833
Adj. R^2	0.66672	0.68823	0.02683	0.02683	−0.09127

Note: Numbers in brackets are standard error. ***, ** and * are significant at the level of 1, 5, and 10%, respectively.

4.2. Analysis of the Time-Delay Impact of High-Speed Rail on County Economic Development

According to the above empirical analysis, high-speed railway opening has a long-term impact on county economic development. Therefore, in this paper, by referring to the event study method proposed by Clarke and Schythe [51], the delay impact of high-speed railway on the county industrial agglomeration level and secondary labor agglomeration degree is explored.

4.2.1. Analysis of the Delay Effect of High-Speed Railway on Industrial Agglomeration in County

High-speed rail opening has a negative impact on the industrial agglomeration level (sec) in counties to a certain extent. However, whether the negative impact has a time lag is verified in this part (see Figure 6). The concrete results show that: (1) From the significant regression coefficient level, high-speed rail opening has a leading effect on and a role in promoting industrial enterprise development. At the same time, high-speed rail opening has obvious delay effect on the county industry's agglomeration and development. When the lag period is the second phase (i.e., the second year), a county's industrial agglomeration level show obvious correlation with the high-speed rail opening, thus indicating that a county's high-speed rail opening will have a long-term negative impact on the industrial agglomeration. (2) From the interaction term's regression coefficient, the leading period is the promoting effect, while the lagging period is the negative effect. The possible reason is that Chinese counties' development levels are relatively behind those of others at present, and that high-speed rail opening has intensified the siphon effect of central cities.

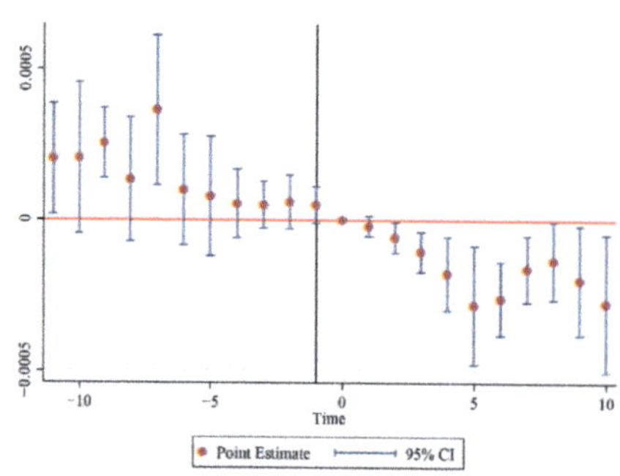

Figure 6. Regression results of the delayed impact of high-speed railway opening on industrial agglomeration in counties. Note: Since there was only one high-speed train in 2003 to 2008, the lag time was only 10 years in 2019 minus 2008.

4.2.2. Analysis of High-Speed Railway's Time-Delay Impact on Counties' Secondary Industry Labor Force Agglomeration

High-speed railway opening promotes secondary labor force agglomeration (emp) to high-speed railway counties. This part continues to explore the delay of high-speed railway to secondary labor force agglomeration in counties and verifies its dynamic lag (see Figure 7). The results show that: In Section 4.1.2, which examines the impact of secondary labor force agglomeration in county units in the year of high-speed railway opening, we discovered that the introduction of high-speed railway encouraged secondary labor force agglomeration. However, we find that it is not the case from the dynamic time-delay analysis of the secondary labor force agglomeration. The influence of high-speed railway on the agglomeration of secondary labor force in county units reveals an erratic state. The elimination of China's demographic dividend and the restriction of the household registration system in the panel data analysis from 2003 to 2019 may be the cause, which diminished the influence of high-speed rail on secondary labor flow.

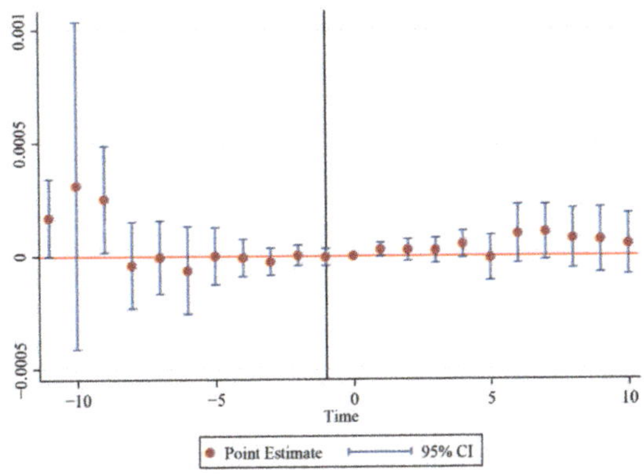

Figure 7. Regression results of the delayed impact of high-speed railway opening on secondary industry labor force agglomeration in counties. Note: Since there was only one high-speed train in 2003 to 2008, the lag time was only 10 years in 2019 minus 2008.

5. Robustness Test

5.1. Endogenic Processing

Excluding endogenous problems' influence is a prerequisite for DID analysis. We mainly referred to Huber and Steinmayr [52] and Jun Zhang [53] by investigating counties with less discourse power and higher randomness as research subjects; two major measures of PSM matching analysis were carried out on the experimental and control groups of the county units to ensure the selection of the county units met the three measures of the randomness hypothesis test, processing the matching of other relevant characteristics of samples, and carrying out a collinearity test of the variables to cause the experimental results to be more reliable. The specific measures are as follows:

1. The inauguration of high-speed rail must be a quasi-natural experiment to pass the random hypothesis test. This is the basis of applying the multi-phase DID model analysis. In his article on high-speed rail's effect on county economic development, Jun Zhang noted the close connection between the planning and construction of high-speed rail lines and stations and the degree of regional economic development, as well as their geographic location and other factors. The likelihood that a central city will open high-speed rail increases with its economic power and political clout. Given this, we opted not to use the prefecture-level cities that many academics have chosen to analyze to examine high-speed rail's effects on urban economic development. To address the endogenous issue, we instead selected counties with lower discourse power and higher randomness as the research subjects.

2. To further limit endogeneity, a multi-phase PSM-DID model based on county unit matching was created in this study. The population, level of economic development, and market potential in county units vary greatly, and this unit heterogeneity causes a clear bias in policy effect estimation. Up to a point, the sample selectivity bias could be reduced by matching the propensity scores of the experimental and control groups based on the characteristics of county units.
3. To test whether there was collinearity among the variables, the collinearity diagnosis was conducted (Table 5). Except for the gross regional product, the VIF value was no more than 6, which proved that there was no collinearity among the variables or that there was to a lesser extent.

Table 5. Collinearity diagnostics table.

	Model 1	Tolerance	VIF	Model 2	Tolerance	VIF
dependent variable	sec			emp		
independent variable	emp	0.325	3.079	sec	0.422	2.368
	lnpop	0.326	3.066	lnpop	0.318	3.145
	lngdp	0.094	10.687	lngdp	0.085	11.832
	lnpot	0.318	3.144	lnpot	0.382	2.621
	lnlopot	0.173	5.786	lnlopot	0.173	5.788
	lnfid	0.252	3.968	lnfid	0.249	4.021

5.2. Parallel Trend Test

The DID model's basic assumption is that the treatment and the control groups have parallel trends; that is, the two county unit groups should have the same trend before the high-speed railway opening. To test the robustness of high-speed rail opening's influence, this paper adopts the method of changing the window width before and after the county high-speed rail opening for verification. The specific formula was as follows:

$$Y_{it} = \beta_0 + \delta_1 \sum_{k\geq 3}^{-2} city_i * post_{i,tc0+k} + \beta_2 xlist_{it} + \mu_i + \lambda_t + \varepsilon_{it} \quad (2)$$

where $tc0$ represents the year of the high-speed railway opening, and $t - tc0 = k = -3$, -2, -1, 0, 1, 2, 2, 3 represent the county units set dummy variables for three years, two years, one year, one year, two years, and three years before the high-speed railway opening, and construct the interaction terms between the corresponding county units and the time dummy variables. Parallel trend tests were carried out on the industrial agglomeration level (sec) and the secondary labor agglomeration degree (emp) of the county units, respectively (see Table 6 and Figure 8). The results show that: ① From the regression coefficient, the regression coefficients of the industrial agglomeration level and secondary labor agglomeration degree are not significant in the three years before the high-speed railway opening, while the baseline group (to completely exclude the collinearity problem, the first period before the policy is usually selected as the baseline group) is significant in the later years. The results indicated that the county units with and without high-speed rail services had the same time trend at least three years before the high-speed rail service. The parallel trend test results verified the multi-period PSM-DID regression results' robustness. ② The parallel trend test chart shows the dynamic economic effects between different years under the high-speed railway policy's impact. The chart indicates that the interaction term's coefficients were not significantly different from 0 before the high-speed railway opening, and that the confidence intervals all contain 0 values, indicating that there was no significant difference between the county units of the experimental and control groups before the high-speed railway opening, which satisfies the hypothesis of the parallel trend. In addition, the parallel test trends of sec and emp after implementing the policy were opposite, thus indicating that the high-speed rail opening significantly differed in its impact on the industrial development and secondary labor force agglomeration across counties.

Table 6. Parallel trend test table of industrial agglomeration level and secondary labor agglomeration degree in county units.

	(1)	(2)
Variables	**Sec**	**Emp**
pre_3	$1.59 \div 10^5$	$-1.41 \div 10^5$
	$(3.57 \div 10^5)$	$(3.32 \div 10^5)$
pre_2	$-3.47 \div 10^5$	$-1.38 \div 10^5$
	$(3.00 \div 10^5)$	$(1.85 \div 10^5)$
current	$-3.23 \div 10^5$	$1.11 \div 10^5$
	$(2.64 \div 10^5)$	$(1.80 \div 10^5)$
post_1	$-4.69 \div 10^5$	$3.53 \div 10^5$
	$(3.69 \div 10^5)$	$(2.33 \div 10^5)$
post_2	$-8.24 \div 10^5$ **	$3.30 \div 10^5$
	$(4.15 \div 10^5)$	$(3.07 \div 10^5)$
post_3	−0.000112 **	$4.09 \div 10^5$
	$(4.42 \div 10^5)$	$(3.35 \div 10^5)$
post_4	−0.000168 **	$6.99 \div 10^5$ **
	$(7.73 \div 10^5)$	$(3.41 \div 10^5)$
post_5	−0.000174 **	$9.92 \div 10^5$ *
	$(7.19 \div 10^5)$	$(5.61 \div 10^5)$
lnpop	$-2.95 \div 10^5$	$5.54 \div 10^5$
	$(7.44 \div 10^5)$	$(7.04 \div 10^5)$
lngdp	0.000697 ***	$8.59 \div 10^5$
	(0.000127)	$(6.40 \div 10^5)$
lnpot	−0.000467 ***	−0.000251 **
	(0.000152)	(0.000107)
lnlopot	$6.47 \div 10^5$	0.000123
	(0.000163)	(0.000117)
lnfid	$-4.97 \div 10^5$ **	$-3.12 \div 10^5$
	$(2.12 \div 10^5)$	$(1.91 \div 10^5)$
Constant	−0.00914 ***	−0.00146
	(0.00161)	(0.00128)
Observations	810	810
R-squared	0.968	0.967

Note: Numbers in brackets are standard errors. ***, ** and * are significant at the level of 1, 5, and 10%, respectively.

5.3. Placebo Test

There were individual differences over time between the county units, regardless of whether high-speed rail was opened, which led to systematic bias in the regression results. The discrepancies in the industrial concentration level (sec) and secondary labor concentration degree (emp) between the two groups of county units should not change over time if the high-speed rail development did not occur. The discrepancies in the industrial concentration level (sec) and secondary labor concentration degree (emp) between the two county unit groups should not change over time if high-speed rail development did not occur. The discrepancies in the industrial concentration level (sec) and secondary labor concentration degree (emp) between the two groups of county units should not change over time if high-speed rail development did not occur. To verify the regression results' reliability, Liu Ruiming et al. (2020) employed a method that is extensively used in the placebo test to randomly produce group tests in the research [54]. The kernel density coefficient and *p*-value scatter plot of the virtual policy variables were compared and examined to see whether there were any significant differences between them and the actual values by randomly generating the processing groups of counties that opened high-speed rail and repeating the regression 1000 times (see Figure 9). As seen from the figure, both the industrial agglomeration level (sec) and the concentration degree of the secondary industry labor force (emp) indicate that, in the case of random sampling, the regression coefficients close to the real benchmark (solid line) tsec = −0.0000457 and temp = 0.0000199 were all small probability events, thus suggesting that high-speed rail opening affected the regional economy. The aforementioned empirical investigation results are reliable and the placebo test was passed.

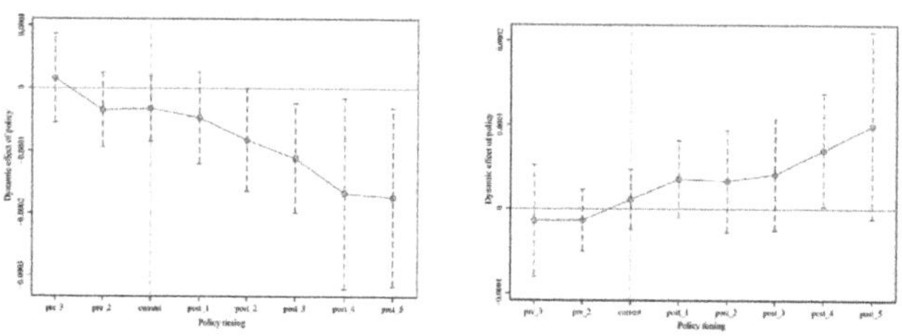

(a) (b)

Figure 8. Parallel trend test diagram of industrial agglomeration level and secondary labor agglomeration degree in county units. (**a**) Test chart of parallel trend of industrial agglomeration level. (**b**) Test chart of parallel trend of concentration degree of secondary industry labor force.

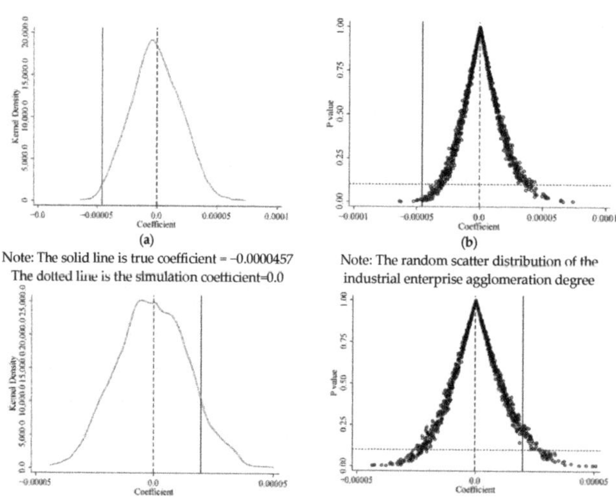

Figure 9. Placebo test chart of industrial agglomeration level and secondary labor agglomeration degree in county units. (**a**) Kernel density coefficient diagram of the industrial concentration level. (**b**) Scatter plot of *p*-value of placebo test for the industrial concentration level. (**c**) Kernel density coefficient diagram of the concentration degree of secondary labor force. (**d**) Scatter plot of *p*-value of placebo test for the concentration degree of secondary labor force.

6. Research Conclusions and Policy Implications

High-speed rail opening eases the historically severely imbalanced regional economy growth brought on by China's high flow cost of labor and other generating factors. The key to the balanced and high-quality development of China's economy in the future is the transformation and upgrading of county industrial and demographic structures. In the context of China's rapid urbanization and regional development transformation, the multi-mode transportation system, including airports, high-speed railways, rail transit, expressways, and buses, has developed into a crucial foundation for coping with the "congestion costs" of big cities, such as urban industrialization and rising housing prices. Many studies and statistical evidence have demonstrated how developing a high-speed rail network has aided in the trend toward suburbanization, regional integration, and cross-regional travel. For instance, Xiongbin Lin and Yuan Lu discovered, through a questionnaire study, that 7% of commuters use the Beijing–Tianjin cross-city high-speed railway [55]. This study examines the new route of county transformation and upgrading in our nation regarding high-speed rail's impact on the county industrial and secondary labor agglomeration. The study views the launch of high-speed rail as a sort of natural experiment. A multi-period PSM-DID model was built to empirically investigate the effects of introducing high-speed rail on the concentration of industrial firms and the secondary labor force in the counties using panel data from 1791 non-central urban units in China from 2003 to 2019.

The findings indicated that introducing high-speed rail changed counties' economic landscapes, had an adverse effect on industrial agglomeration, and had a favorable effect on the secondary labor concentration. According to the examination of high-speed rail's time delay, industrial firm development in county regions is normally negatively impacted by high-speed rail for a long period before turning around eight years later. High-speed rail has a long-term positive effect on the concentration of the secondary labor force, but does not show an obvious trend of orderly concentration. Here, we found China's high-speed rail counties are currently developing primarily into satellite cities formed by population agglomeration, or "sleeping cities". Developing a high-speed rail network has, according to pertinent studies, accelerated the national gradient transfer of low-end manufacturing businesses to less-developed regions [56–59]. The "siphon effect" of center cities prevents industrial units developing in neighboring cities; however, on the small scale of central cities and surrounding counties, the impact of intercity high-speed rail construction on neighboring county development is minimal. Additionally, we investigated how the potential of the national and local markets affected the county's economic growth. We discovered that, while the local market potential attracted a secondary labor force concentration, the national level mainly had a negative impact on the county's industrial development.

High-speed rail's effect on the county economic development was examined using a theoretical and an econometric model. Owing to the restricted research capacity and data collection, the research method still has major flaws that need to be addressed, notably the following two main points: (1) The theoretical framework and model need to be further refined. Cities are intricate systems. The variability of numerous significant economic determinants and unique county development features is still ignored by the theoretical framework of the interaction between county economic development and the high-speed rail network that the study created. (2) The data processing and characterization variable selection need to be further studied. Many county units with missing data were eliminated because of the small number of county samples and the lengthy period (2003–2019). The research results' accuracy differed from the real results due to a linear interpolation approach being employed to supplement a limited number of county units with missing data. The data availability also restricted the choice of variables. We propose that, in the future, research into the scope and interactions of multivariate factors, the data's veracity, and the use of high-speed rail for mass transit in the metropolitan region be performed in greater depth.

Author Contributions: Conceptualization, F.C. and H.H.; methodology, F.C. and H.H.; software, F.C.; validation, F.C.; formal analysis, F.C.; investigation, F.C.; resources, H.H.; data curation, F.C.; writing—original draft preparation, F.C.; writing—review and editing, F.C. and H.H.; visualization, F.C.; supervision, H.H.; project administration, H.H.; funding acquisition, H.H. All authors have read and agreed to the published version of the manuscript.

Funding: This research was funded by the Center for Balance Architecture of Zhejiang University (Project No: K Heng 20203512-02B, Index and planning methods of resilient cities).

Data Availability Statement: The data presented in this study are available upon request from the authors.

Conflicts of Interest: The authors declare no conflict of interest.

Appendix A

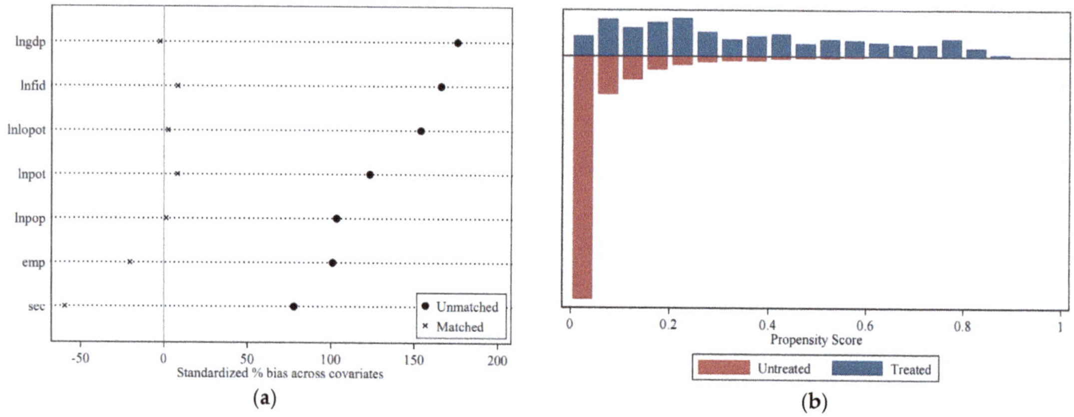

Figure A1. PSM balancing table. (**a**) PSM equilibrium test diagram. (**b**) P-score balance test chart.

Table A1. PSM-nearest-neighbor matching result.

Citypost	Coefficient	Std. Err.	z	$p > \|z\|$	[95% Conf.	[Interval]
lnpop	−0.0404	0.1414	−0.2900	0.7750	−0.3175	0.2367
lngdp	1.5066	0.2019	7.46000	0.0000	1.1108	1.9024
lnpot	−0.1788	0.3579	−0.5000	0.6170	−0.8802	0.5226
lnlopot	0.0463	0.0976	0.4700	0.6350	−0.14496	0.2376
lnfid	0.4051	0.1035	3.9100	0.0000	0.2022	0.6079
_cons	−29.4528	2.3367	−12.6000	0.0000	−34.0327	−24.8729

Table A2. PSM-nearest-neighbor matching result.

Variable	Sample	Treated	Controls	Difference	S.E.	T-Stat
sec	Unmatched	0.0007	0.0002	0.0005	0.0000	16.4200
	ATT	0.0007	0.0011	−0.0004	0.0001	−5.6500
	ATU	0.0003	0.0002	−0.0001	.	.
	ATE	.	.	−0.0001	.	.
emp	Unmatched	0.0008	0.0003	0.0005	0.0000	21.5500
	ATT	0.0008	0.0010	−0.0001	0.0001	−2.1300
	ATU	0.0003	0.0002	−0.0001	.	.
	ATE	.	.	−0.0001	.	.
	Untreated	3373	3373			
	Treated	333	333			
	Total	3706	3706			

Table A3. Descriptive statistics.

Variable	Obs	Mean	Std. Dev.	Min	Max
id	3706	109.5	62.939	1	218
year	3706	2011	4.9	2003	2019
sec	3706	0	0.001	0	0.004
emp	3706	0	0	0	0.003
hsr	3706	498.839	869.451	0	2019
citypost	3706	0.09	0.286	0	1
lnpop	3706	3.469	0.913	0.693	5.094
lngdp	3706	13.318	1.305	8.971	16.759
lnpot	3706	0.162	0.25	0	3.321
lnlopot	3706	8.228	1.827	2.499	14.463
lnfid	3706	12.678	1.568	4.159	28.597
tmp	3706	0.503	0.287	0	1
pscore	3706	0.09	0.154	0	0.984
treated	3706	0.09	0.286	0	1
support	3706	1	0	1	1
weight	833	4.449	32.598	0.333	538.667

References

1. Donaldson, D. Railroads of the Raj: Estimating the Impact of Transportation Infrastructure. *Am. Econ. Rev.* **2018**, *108*, 899–934. [CrossRef]
2. Garrison, W.L.; Souleyrette, R.R. Transportation, Innovation, and Development: The Companion Innovation Hypothesis. *Logist. Transp. Rev.* **1996**, *32*, 5–38.
3. Givoni, M. Development and Impact of the Modern High-Speed Train: A Review. *Transp. Rev.* **2006**, *26*, 593–611. [CrossRef]
4. Duranton, G.; Turner, M.A. Urban Growth and Transportation. *Rev. Econ. Stud.* **2012**, *79*, 1407–1440. [CrossRef]
5. Sasaki, K.; Ohashi, T.; Ando, A. High-Speed Rail Transit Impact on Regional Systems: Does the Shinkansen Contribute to Dispersion? *Ann. Reg. Sci.* **1997**, *31*, 77–98. [CrossRef]
6. Shao, S.; Tian, Z.; Yang, L. High Speed Rail and Urban Service Industry Agglomeration: Evidence from China's Yangtze River Delta Region. *J. Transp. Geogr.* **2017**, *64*, 174–183. [CrossRef]
7. Zheng, S.; Kahn, M.E. China's Bullet Trains Facilitate Market Integration and Mitigate the Cost of Megacity Growth. *Proc. Natl. Acad. Sci. USA* **2013**, *110*, E1248–E1253. [CrossRef]
8. Heuermann, D.F.; Schmieder, J.F. The Effect of Infrastructure on Worker Mobility: Evidence from High-Speed Rail Expansion in Germany. *J. Econ. Geogr.* **2019**, *19*, 335–372. [CrossRef]
9. Verma, A.; Sudhira, H.S.; Rathi, S.; King, R.; Dash, N. Sustainable Urbanization Using High Speed Rail (HSR) in Karnataka, India. *Res. Transp. Econ.* **2013**, *38*, 67–77. [CrossRef]
10. Wu, S.; Lee, J. Analysis of Core Competencies for Overseas Expansion of China's High-Speed Railway Industry. *J. Korean Soc. Railw.* **2020**, *23*, 374–383. [CrossRef]
11. Okamoto, C.; Sato, Y. Impacts of High-Speed Rail Construction on Land Prices in Urban Agglomerations: Evidence from Kyushu in Japan. *J. Asian Econ.* **2021**, *76*, 101364. [CrossRef]
12. Tombe, T.; Zhu, X. Trade, Migration, and Productivity: A Quantitative Analysis of China. *Am. Econ. Rev.* **2019**, *109*, 1843–1872. [CrossRef]
13. Elhorst, J.P. A Spatiotemporal Analysis of Aggregate Labour Force Behaviour by Sex and Age across the European Union. *J. Geogr. Syst.* **2008**, *10*, 167–190. [CrossRef]
14. Dong, X.; Zheng, S.; Kahn, M.E. The Role of Transportation Speed in Facilitating High Skilled Teamwork across Cities. *J. Urban Econ.* **2020**, *115*, 103212. [CrossRef]
15. Hall, P. Magic Carpets and Seamless Webs: Opportunities and Constraints for High-Speed Trains in Europe. *Built Environ.* **2009**, *35*, 59–69. [CrossRef]
16. Mayer, H. High-Speed Rail Passenger Transportation and Regional Development in Midwest. *Traffic Q.* **1967**, *21*, 395–405.
17. Sands, B.D. *The Developmental Effects of High-Speed Rail Stations and Implications for California*; Working Paper; Alexandrine Press: Abingdon, UK, 1993; 59p.
18. Blum, U.; Haynes, K.; Karlsson, C. Introduction to the Special Issue The Regional and Urban Effects of High-Speed Trains. *Ann. Reg. Sci.* **1997**, *31*, 1–20. [CrossRef]
19. Chen, C.L.; Hall, P. The Impacts of High-Speed Trains on British Economic Geography: A Study of the UK's InterCity 125/225 and Its Effects. *J. Transp. Geogr.* **2011**, *19*, 689–704. [CrossRef]
20. Willigers, J.; Wee, B.V. High-Speed Rail and Office Location Choices. A Stated Choice Experiment for the Netherlands. *J. Transp. Geogr.* **2011**, *19*, 745–754. [CrossRef]

21. Diao, M. Does Growth Follow the Rail? The Potential Impact of High-Speed Rail on the Economic Geography of China. *Transp. Res. Part Policy Pract.* **2018**, *113*, 279–290. [CrossRef]
22. Krugman, P. Increasing Returns and Economic Geography. *J. Polit. Econ.* **1991**, *99*, 483–499. [CrossRef]
23. Faber, B. Trade Integration, Market Size, and Industrialization: Evidence from China's National Trunk Highway System. *Rev. Econ. Stud.* **2014**, *81*, 1046–1070. [CrossRef]
24. Du, X.Q.; Peng, M.W. Do High-Speed Trains Motivate the Flow of Corporate Highly Educated Talents? *Bus. Manag. J.* **2017**, *39*, 89–107. [CrossRef]
25. Zingales, L. In Search of New Foundations. *J. Finance* **2000**, *55*, 1623–1653. [CrossRef]
26. Meng, X.; Lin, S.; Ren, G. The Effect of HSR Connection on Urban Cluster Centrality: Evidence from China. *Appl. Econ.* **2022**, *54*, 3088–3102. [CrossRef]
27. Rosenthal, S.S.; Strange, W.C. The Attenuation of Human Capital Spillovers. *J. Urban Econ.* **2008**, *64*, 373–389. [CrossRef]
28. Guirao, B.; Luis Campa, J.; Casado-Sanz, N. Labour Mobility between Cities and Metropolitan Integration: The Role of High Speed Rail Commuting in Spain. *Cities* **2018**, *78*, 140–154. [CrossRef]
29. Chen, D.; Ma, Y.; Martin, X.; Michaely, R. On the Fast Track: Information Acquisition Costs and Information Production. *J. Financ. Econ.* **2022**, *143*, 794–823. [CrossRef]
30. Zheng, L.; Chang, Z.; Martinez, A.G. High-Speed Rail, Market Access, and the Rise of Consumer Cities: Evidence from China. *Transp. Res. Part Policy Pract.* **2022**, *165*, 454–470. [CrossRef]
31. Wu, Q.B.; Tao, J.J.; Wu, K.; Zhao, M.X. Daily Commuting Behavor Characteristics of High Speed Rail Passengers in the Yangtze River Delta Region: A Case Study of Shanghal-Hangzhou, Nanjing-Hangzhou, and Hangzhou-Ningbo Lines. *City Plan. Rev.* **2018**, *42*, 90–97+122.
32. Duranton, G. Urban Evolutions: The Fast, the Slow, and the Still. *Am. Econ. Rev.* **2007**, *97*, 197–221. [CrossRef]
33. Duranton, G.; Puga, D. Nursery Cities: Urban Diversity, Process Innovation, and the Life Cycle of Products. *Am. Econ. Rev.* **2001**, *91*, 1454–1477. [CrossRef]
34. GLA Economics. *London's Creative Sector: 2004 Update*; GLA Economics: London, UK, 2007.
35. Kim, K.S. High-Speed Rail Developments and Spatial Restructuring—A Case Study of the Capital Region in South Korea. *Cities* **2000**, *17*, 251–262. [CrossRef]
36. Goodman-Bacon, A. Difference-in-Differences with Variation in Treatment Timing. *J. Econom.* **2021**, *225*, 254–277. [CrossRef]
37. Callaway, B.; Sant'Anna, P.H.C. Difference-in-Differences with Multiple Time Periods. *J. Econom.* **2021**, *225*, 200–230. [CrossRef]
38. Ahlfeldt, G.M.; Feddersen, A. From Periphery to Core: Measuring Agglomeration Effects Using High-Speed Rail. *Serc Discuss. Pap.* **2017**, *18*, 355–390. [CrossRef]
39. Redding, S.J.; Turner, M.A. Transportation Costs and the Spatial Organization of Economic Activity. *Handb. Reg. Urban Econ.* **2015**, *5*, 1339–1398. [CrossRef]
40. Wen, M. Relocation and Agglomeration of Chinese Industry. *J. Dev. Econ.* **2004**, *73*, 329–347. [CrossRef]
41. Jin, Y.; Chen, Z.; Lu, M. Regional Industrial Agglomeration in China: Economic Geography, New Economic Geography and Economic Policy. *Econ. Res. J.* **2006**, *4*, 79–89.
42. Nakagawa, D.; Hatoko, M. Reevaluation of Japanese High-Speed Rail Construction. *Transp. Policy* **2007**, *14*, 150–164. [CrossRef]
43. Wang, F.; Wei, X.; Liu, J.; He, L.; Gao, M. Impact of High-Speed Rail on Population Mobility and Urbanisation: A Case Study on Yangtze River Delta Urban Agglomeration, China. *Transp. Res. Part Policy Pract.* **2019**, *127*, 99–114. [CrossRef]
44. Song, C.-X.; Qiao, C.-X.; Luo, J. Does High-Speed Rail Opening Affect the Health Care Environment?–Evidence From China. *Front. Public Health* **2021**, *9*, 708527. [CrossRef] [PubMed]
45. Wang, G.D.; Niu, Y.; Sun, F.; Wang, K.Y.; Qian, J.; Li, F. Evolution and spatial characteristics of tourism field strength of cities linked by high-speed rail (HSR) network in China. *J. Geogr. Sci.* **2017**, *27*, 835–856. [CrossRef]
46. Shaw, S.-L.; Fang, Z.; Lu, S.; Tao, R. Impacts of High Speed Rail on Railroad Network Accessibility in China. *J. Transp. Geogr.* **2014**, *40*, 112–122. [CrossRef]
47. Jiao, J.; Wang, J.; Jin, F.; Dunford, M. Impacts on Accessibility of China's Present and Future HSR Network. *J. Transp. Geogr.* **2014**, *40*, 123–132. [CrossRef]
48. Ke, X.; Chen, H.; Hong, Y.; Hsiao, C. Do China's High-Speed-Rail Projects Promote Local Economy?—New Evidence from a Panel Data Approach. *China Econ. Rev.* **2017**, *44*, 203–226. [CrossRef]
49. Deng, T.; Wang, D.; Yang, Y.; Yang, H. Shrinking Cities in Growing China: Did High Speed Rail Further Aggravate Urban Shrinkage? *Cities* **2019**, *86*, 210–219. [CrossRef]
50. Qin, Y. "No County Left behind?" The Distributional Impact of High-Speed Rail Upgrades in China. *J. Econ. Geogr.* **2017**, *17*, 489–520. [CrossRef]
51. Clarke, D.; Schythe, K.T. EVENTDD: Stata Module to Panel Event Study Models and Generate Event Study Plots. *Stat. Softw. Compon.* **2021**.
52. Huber, M.; Steinmayr, A. A Framework for Separating Individual-Level Treatment Effects From Spillover Effects. *J. Bus. Econ. Stat.* **2021**, *39*, 422–436. [CrossRef]
53. Zhang, J. High-Speed Rail Construction And County Economic Development: A Study Based on Satellite Light Data. *China Econ. Q.* **2017**, *16*, 1533–1562. [CrossRef]

54. Liu, R.M.; Mao, Y.; Kang, Y.K. Deregulation, Market Vitality and Tourism Economy Development: Evidence from Chinese Cultural System Reform. *Econ. Res. J.* **2020**, *55*, 115–131.
55. Sheng, J. Research on the Relevance between Land-based High-Speed Traffic Trunk and the Rank-sizeDistribution of City in China. Master's Thesis, Zhejiang University, Hangzhou, China, 2016.
56. Deng, H.H.; Yang, L.X.; Pang, X.T. Whether the opening of high-speed rail can help upgrade industrial structure: Facts and Mechanisms. *J. Financ. Econ.* **2020**, *46*, 34–48. [CrossRef]
57. Jiang, H.X.; Cai, H.Y.; Meng, X.C. The Structural Impact of High Speed Rail Upon Urban Industries in China. *Hum. Geogr.* **2017**, *32*, 132–138. [CrossRef]
58. Qiao, B.; Zhang, X.; Lei, C. The Effect of High-Speed Railway, Agglomeration of Productive Service Industry and Upgrading of Manufacturing Industry. *Econ. Rev.* **2019**, 80–96. [CrossRef]
59. Zhu, W.T. Impact of high-speed rail service supply on spatial agglomeration of provincial manufacturing industry. *Ind. Econ. Res.* **2019**, 27–39. [CrossRef]

land

MDPI

Article

Tourism-Led Change of the City Centre

José Fernandes [1,*], Pedro Chamusca [2], Rubén Lois [3], Helena Madureira [1], Juliano Mattos [1] and Jorge Pinto [4]

1 Geography and Spatial Planning Research Centre (CEGOT), University of Porto, 4150-564 Porto, Portugal; hmadureira@letras.up.pt (H.M.); up200704959@up.pt (J.M.)
2 Communication and Society Research Centre (CECS), University of Minho, 4710-057 Braga, Portugal; pedrochamusca@ics.uminho.pt
3 ANTE—Grupo de Investigación de Análisis Territorial, Universidad de Santiago de Compostela, 15705 Santiago de Compostela, Spain; rubencamilo.lois@usc.es
4 Geography and Spatial Planning Research Centre (CEGOT), ISCET, 4050-180 Porto, Portugal; jpinto@iscet.pt
* Correspondence: joseriof@letras.up.pt

Abstract: In multicentric and increasingly complex urban regions, a city centre reinvents itself. In the case of Porto, tourism was essential for its "Baixa" renaissance. A relevant increase in visitors meant also a dramatic increase in real estate prices and significant land-use change. In field interviews, retailers noticed a "new life" before COVID-19 arrived, remarking on the positive role of tourism on urban rehabilitation and the economic viability of companies, and the negative effects for residents and traditional shops, directed to the common resident. In this article, we present and discuss its main effects in this exceptional area in Portugal's second city. We also discuss tourism dependency and the challenge of sustainability in a high-density context, defending public policies oriented for a "city with tourists" that replaces the current construction of a "city of tourists".

Keywords: urban tourism; overtourism; gentrification; land use; sustainability; public policies

1. Introduction

In recent decades, mutations in the city centre have been particularly intense as low-cost flights have become popular and changing places is easier (specially for countries within the eurozone and under the Maastricht treaty), while digital platforms facilitate travel and accommodation. As a consequence, especially historical, cultural, and environmental attractive old cities appeal to visitors and all kinds of city users. This new flowing population and real estate investment are associated with urban renewal and with soaring prices, triggering functional and residential gentrification, a term coined by Glass [1] that relates to the process of transformation of marginalized and/or traditional neighbourhoods for wealthier and more sophisticated solutions, with the replacement of residents and activities [2,3].

It is common that a process of beautification occurs, as the image of the city becomes more relevant, more so in the central squares and the façades of the most visited streets. As a consequence of social change and the process of rising prices, economic land uses also suffer significative alterations.

The main objective of this article is to discuss the change of the city centre in the face of dependence from tourism, considering "sustainable tourism" and "sustainable urbanism" principles. For that, we use Porto as a case study. And we do that for good reasons. The second city of Portugal was elected European Best Destination in 2012, 2014 and 2017, and the European Best City-Break Destination in 2020. The number of overnight stays went from 2,102,481 to 4,819,168 from 2013 to 2022 (129.2%) whereas in Lisbon the change was 78.9%, and there were 6170 Airbnb units per 100.000 inhabitants prior to COVID, in 2018, which is higher than Paris and Barcelona [4]. Also, it was in Porto that, in July 2021, 17 European cities (including Venice and Prague) and some companies (including Airbnb) signed a document declaring their concern with excessive tourism and, in the words of the local authorities, decided to "align tourism with the best practices of sustainable policies".

To do so, our research seeks to shed light on three research questions:

- How is tourism conditioning the city centre's social and economic evolution?
- How are tourism and public policies perceived by economic agents?
- How should urbanism for sustainability be designed and implemented in a tourism-dependent city centre?

The analysis follows the following structure: In Section 2, we deal with the theoretical framework of the research. To do so, we consider the main dynamics associated with tourism in city centres and recent evolution of urban public policies in selected European city centres. Section 3 presents the materials and methods, as well as the city of Porto. Section 4 present the results for Porto city centre, and in Section 5, there is a discussion of the effects and dynamics of tourism on its relation with a sustainable approach. Finally, Section 6 presents the main conclusions and their relation to the paper's theoretical framework and research questions.

2. City Centre, Sustainability, and Tourism

2.1. Centre and Centrality: Recent Transformations

A city may have a centre or several centres, depending on its history and dimension, and the combination of the conditions for centrality (land value, symbolic value, and accessibility condition) is different in each place. Central areas differ, from the old, traditional city centre, to the new areas of the European city with high buildings which host multinationals and large companies, or new sexy and gentrified areas where culture and recreation is relevant [5,6].

Undoubtedly, the symbolic dimension of the city centre is essential and allows it to be set apart from other locations, as well as to distinguish the central city from the expanded, polycentric, and fragmented urban area. This is where major public actors (government offices, palaces, and historic universities) and large private companies (banks, insurance companies, specialized occupations offices, and luxury trade) find their place. All these activities value the city's central heritage and the highlighted examples of civil architecture of the last two or three hundred years. But heritage also bears historical significance conformed by monuments and other older buildings. That is concentrated in another type of centre, the so-called "historical" centre, since the condition of centrality is normally related here with a long period of time and the capacity the city had to retain essential elements of its construction and reconstruction over the centuries [7,8].

Historical centres and economical centres are the most attractive areas for urban tourism. The massification of access to air travel, and more economic capacity and time also means a higher value of aesthetic and cultural experiences. As a consequence, city-breaks compete with locations of 20th-century massification (sun and beach destinations), and extend far beyond the usual urban venues of Paris, London, Venice, Rome and Barcelona. As a result, in several cities, and especially in their centre, accommodation, restaurants and other tourism-oriented facilities increase dramatically.

In addition, with the emergence of the so-called "collaborative economy", best known as the "platform economy" [9,10], Airbnb and others have encouraged the multiplication of apartments and flats for rent [11,12]. These short-term rentals have helped to occupy several of the residential places that still existed in the centre of cities such as Barcelona, Lisbon, Porto, or Palma de Mallorca [13–16] and have denied about half to three quarters of the offers of new housing in New York [17]. If tourism and short-rent locations created the opportunity to invest in old housing, improving living conditions in decadent buildings, they also gave new meaning to competition in the centre regarding land uses, as it became a place where the presence of the floating population is more important than the "common population", and the price for housing, products, meals, and services become impossible for a good part of people to live and retailers to resist.

It is within this context that the rent gap applies, as the difference between the rent realized from a plot of land and the potential rent if it were developed to its "highest and best" use increases and attracts real estate interest, including international. Simultaneously, overtourism emerges as the tourist densification of specific streets and squares, occurs, along with the proliferation of hotel concentrations, forcing the displacement of the more fragile residents and activities [18,19].

2.2. Sustainable Urban Tourism

The relation between sustainability and tourism is based on several processes and debates, including the questions posed by the "limits to growth" concept [20,21], or the increasing perception of the negative outcomes of tourism growth in destination areas [22]. Definitions of sustainable tourism normally embody a holistic perspective, incorporating a suitable balance between economic, sociocultural and environmental aspects in long-term development perspectives [23].

Paradoxically, although the debate on sustainable tourism has been widely embraced by policy makers from both public and private organisations (although, in many occasions, some measures have been denounced as mere "green washing" exercises, by masquerading as ecological discourse for their own benefit in reducing operating costs or to keep in line with the ideological change), little to no attention has been given to the conceptualization of sustainability in an urban tourism context [24–26]. And that occurs despite the remarkable growth of urban tourism during the last two decades, the knowledge of its impacts, and the growing number of global, national, and local initiatives taken during the last decades to make cities more sustainable. Several factors may be pointed out to explain this, including the traditional focus of sustainable tourism studies on rural and eco-touristic places or the insufficient research conducted in various aspects of the urban tourism phenomenon itself [25].

Nevertheless, a growing focus on the impacts of what is seen as overtourism has led researchers to engage with sustainable urban tourism issues. The social, economic and environmental impacts of the touristification phenomena have also been increasingly acknowledged by researchers, policy makers, tourism stakeholders and local communities [18,27]. In fact, even though tourism can represent an important opportunity for development, contributing to the creation of employment and improvements in infrastructure, it can also promote the emergence or intensification of urban problems such as those related to the increase in the cost of living, new forms of gentrification and CO_2 emissions, as well as other forms of pollution, some of them related with congestion in circulation.

There is a challenge in achieving the best compromise between the socioeconomical benefits and the socioenvironmental negative impacts of tourism [28]. And despite the lack of a clear definition of the concept of sustainable urban tourism or the ambiguous use of the concepts of overtourism or gentrification (there are no instruments or indexes that will unequivocally define them, nor a correct geographical dimension), there is some convergence on the challenges that sustainable tourism is facing in urban areas, namely the management of conflicts between residents' quality of life and urban development processes associated with the tourism industry, or between the residents' perceived local environmental qualities and the local environmental issues induced by overtourism [24,29].

Many researchers have looked for limits and thresholds as a way to achieve sustainable urban tourism. Based on the existing studies at the time, Saarinen [22] systematised approaches to the limits of tourism growth in three main groups: (i) resource-based limits, related to the carrying capacity model and the search for a limit which cannot be overstepped without serious negative impacts on the available resources; (ii) activity-based limits, related to tourism-centric approaches and the idea that different tourism activities or segments may have different kinds of growth limits; and (iii) community-based limits, aiming to empower specifically the host communities in tourism development. However, there is not a "magic number" [22,30] for the maximum acceptable number of tourists at a destination, a threshold beyond which damage would be created, namely due to the difficulty in evaluating all former dimensions simultaneously. For instance, carrying capacity is not only related to a certain resource but also to human values and perceptions concerning that resource [22] and it is influenced not only by the tourist's behaviours and practices, but also by the environmental and socioeconomical resilience of the destination [30].

The difficulty in defining and applying limit thresholds to sustainable urban tourism development is, in fact, a clear example of the complex nature of the concept of sustainable urban tourism and it mirrors the consequent challenges concerning its application in urban policies, for instance, the challenge of balancing the double-edged nature of tourism [31], as an important economic resource and a generator of negative socioenvironmental impacts. Or the challenge of embracing the multidimensional character of tourism, comprising complex interactions between the industry and the specific urban context.

Even though holistic sustainable urban tourism approaches should encompass both issues inherently related with the source (tourism industry, visitors' behaviours and practices) and issues related with the supply's context-specific characteristics, urban policies are almost limited to intervene, in a more defensive or reactive way, in the urban supply context.

The criticism of the conceptualization and operationalization of sustainable tourism, and particularly of sustainable urban tourism, does not diminish the importance of policies oriented by sustainable urban tourism principles. As it has been highlighted [31], despite the inherent limitations, urban policies may contribute to reduce socioenvironmental problems and to find the best compromises, minimising the negative impacts.

Sustainability has long been a part of urbanism, in different periods in history, much before tourism was important in so many cities [32]. Sustainable urbanism revolves around several pivotal principles aimed at harmonizing the interplay between environmental, socio-cultural, and economic facets in urban contexts. At its core lies the concept of compact and connected cities, emphasizing the creation of walkable neighbourhoods and mixed-use spaces to minimize urban sprawl and foster efficient land utilization as well as proximity, theoretically epitomized on the "15-min city" [33,34], in the sequence of the neighbourhood unit of Clarence Perry and the pedestrian pocket of Peter Calthorpe (see [35]). This model not only reduces reliance on cars but also encourages a sense of community and accessibility to amenities, catering to both residents and tourists.

Another crucial aspect involves prioritizing green infrastructure and biodiversity conservation. Incorporating green spaces and preserving natural habitats within cities bolsters ecological resilience, enhances air quality, and offers recreational havens for urban dwellers and visitors alike [36]. Moreover, sustainable urbanism emphasizes resource efficiency and sustainable mobility by advocating for public transportation, cycling lanes, and pedestrian-friendly pathways. This strategy aims to curb carbon emissions, alleviate traffic congestion, and promote healthier, more sustainable modes of transport [37,38].

Additionally, social inclusivity and community engagement form integral components. Ensuring access to essential services, affordable housing, and public spaces for all residents fosters a more inclusive urban environment. Community involvement in decision-making processes not only amplifies diverse perspectives but also instils a sense of ownership and belonging. These principles, along with resilience, economic prosperity, and innovation, serve as guiding pillars in urban planning and policymaking. By embracing these principles, cities can navigate the complexities of sustainable urban tourism, forging a path toward a more resilient, inclusive, and environmentally conscious urban future where tourism complements rather than compromises the well-being of both locals and visitors.

2.3. *Policies for the City Centre*

In the 1970s and 1980s, many European cities were facing physical decay, and losing both population and economic vitality. In several cases we witnessed a revival of cities and its centres as strategic places for a wide range of projects and dynamics, addressing the economic, architectural, social, cultural, and political dimensions [39–41].

The evolution of urban policies in Europe are for a large part inherited and the result of an ad hoc combination [42]. This is supported by a long-standing autonomy of cities and municipal government, on the one hand, and the strength of states and public policies, on the other. A third aspect has been very relevant in European Union cities more recently, especially those with less investment capacity like Portugal: the impact of the EU funding programmes dedicated to urban areas and policies.

Among the multiple direct ways EU action affects urban policies, it is possible to identify programmes such as RECITE, URBAN, URBACT or INTERREG, financial engineering mechanisms such as JESSICA, research lines in urban themes under Horizon 2020, and initiatives such as the European green capital award, the European mobility week award, the smart cities stakeholder platform, Urban Audit or ICLEI Europe, Agenda 21 and the Covenant of Mayors.

The European urban policy has gone through a winding road of consolidation and maturation [43,44]. Overall, this development reproduces a framework marked by four major challenges: globalization and economic restructuring, resulting in the need to promote a more balanced urban system, with enhanced economic growth and employment; social inclusion and economic restructuring; sustainability; and governance, pointing to the increased capacity of local actors to manage change and provide better response to fiscal, organizational, institutional and administrative problems in a multi-agent and multi-scalar approach, open to public participation.

Within the context of this evolution, the urban sustainability of cities is receiving particular attention since the formalisation of the Cohesion Policy after the Single European Act, as a capital challenge; thus, it is assuming its own characteristic both as a thematic agenda of public policy and as a practical methodology of its territorialization within sustainable urban development.

3. Geographical Context and Methodology

3.1. Baixa

Porto, today, is the main centre of a polycentric urban region which, between the cities of Viana do Castelo and Aveiro, in a 150 km long and 20 km wide stretch, accounts for 3.5 million inhabitants (1/3 of the Portuguese population).

Porto is also the main administrative city of a de facto city made up of several municipalities that reaches almost one million inhabitants in a 10 km radius circle. Therefore, when we talk about the municipality of Porto, and specifically about its Baixa, we are referring to the symbolic centre and the greater geographic concentration of tourism in a vast region.

Being a place of relevant monumental density and the subject of public policies with greater prominence, Porto also plays the role of an institutional centre, having the headquarters of classical organizations such as the chamber of commerce, the regional coordination commission, various regional directorates, and the largest university in the region. In addition, Porto has an airport (Sá Carneiro) that was considered by the Airport Council International to be the best in Europe in 2022 on its category (10–25 million passengers/year), and is the great hub of the hotel industry in the region, where all the information and tourist routes of the Portuguese North are focused.

Porto has greatly increased its expression as a tourist destination. At its symbolic centre, this increase may be considered dramatic, since, 20 years ago, it survived from the effects of intense suburbanization of residence, industry and shopping. At the turn of the century, there were strong investments in the public space of the "old" city centre, in preparation for the celebration of the European Capital of Culture, in 2001. A remarkable improvement in accessibility follows, with the creation of the light rail system, where a line connecting Porto with Gaia (2005), reinforces enormously accessibility in the heart of the city, as it crosses with the east–west line at Trindade station. Low-cost aviation in a city whose historical centre was classified as a World Heritage Site in 1996 has also got to be considered as relevant in the transformation of a decadent centre. The figures of the variation between 2010 and 2019 speak for themselves: number of inbound passengers at the airport +150%; tourist accommodation +293%; traditional accommodation capacity +114%; and overnight stays +171%.

3.2. Methods

In this study, we combine several information sources and procedures. We use bibliography for contextualization, statistical data and other objective references, namely on public policy instruments, and the results of an intensive work of direct collection of quantitative information (with closed responses) and semi-structured interviews of a qualitative nature. We also used data resulting from previous research, such as analyses of the morphology and structure of Porto and on the Airbnb platform and short-term accommodation in the city of Porto and its recent changes [45], updated and now deepened in direct contact, namely retailers and other actors who are simultaneously residents and users of the city centre.

A functional survey conducted every two years, signalling all the changes occurring on the ground floor, was of great use in the perception of recent transformations, with the first and the most recent survey, carried out in 2012 and 2020, respectively, being considered.

An interview was applied in order to collect the local actors' opinion. The interview was held by the authors with the owners or managers of 54 establishments (Figure 1) in November 2021, chosen with attention to economic diversity, geographical distribution, and more tourist-oriented establishments and others, seemingly more oriented to the resident.

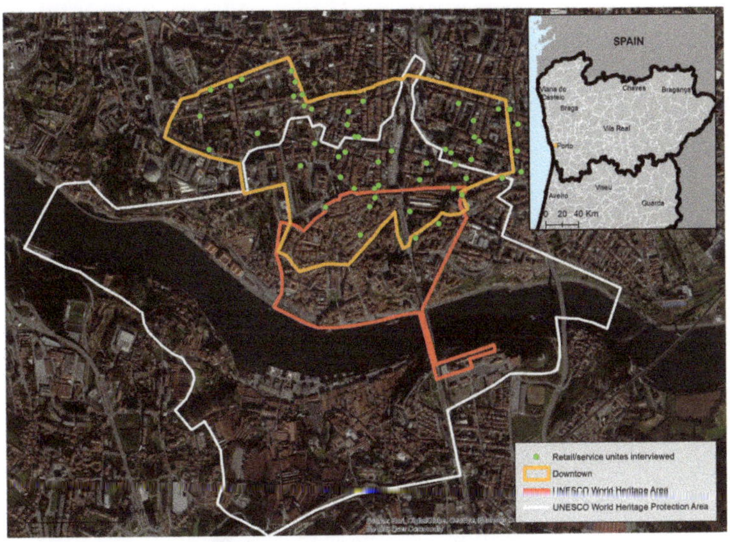

Figure 1. Localization of retail/service units interviewed. Source: Own elaboration.

We considered Baixa as the space already defined as the city of Porto centre (or traditional centre) in other works, keeping in mind the symbolic dimension and accessibility conditions, but, in particular, the concentration of shops and restaurants and the diversity of economic activities.

Regarding the characteristics of the interviewees, among the establishments visited, 44.4% had more than 30 years of existence and 31.5% less than 10 years. The remaining establishments fell within the range of 10 to 30 years of existence. The predominant standing was clearly the average, with only four being considered "luxury" and nine of low standing.

Regarding the diversity of activities, we visited both retail shops and restaurants and other similar units, as well as very few services of a commercial nature. The interviews included 11 shops associated with the sale of articles of personal use, 3 of alimentation articles and 17 "Horeca" units (hotel–restaurant–cafeteria).

With the answers obtained, we designed an initial framework for the characterization of change as perceived by the respondents, of the effectiveness of public policies or, on the contrary, of how processes due to strictly private dynamics were central to triggering the transformation of Baixa.

Semi-structured interviews were used in depth, following the methodological models contrasted by the bibliography. In these interviews, we wanted to know the perceived importance of tourism; the best and worst effects of recent change, and, based on their responses, to draw a predictive scenario for 2030. Finally, interviewees were asked to indicate three projects or measures to be implemented in Baixa. The semi-structured interview seeks to understand the values, opinions, behaviours, and perceptions of the respondents, based on the idea of the "city of citizens".

4. Results

Tourists have been essential for the transformation of land use in Porto city centre, together with other members of an increasing floating population ("digital nomad", Erasmus students, congresspersons. . .). Between July 2012 and July 2020, there was a strong increase in the number of tourist accommodation units (238%), coffee shops and restaurants (46.3%) and non-specialized commerce (3.9%), with a number of souvenir shops and "typical items" (see Figure 2). In 2020, the accommodation establishments (identifiable from the street), the coffee shops and restaurants accounted for 29.7% of the entire offer in Porto's Baixa, while, in 2012, they represented 17.4%.

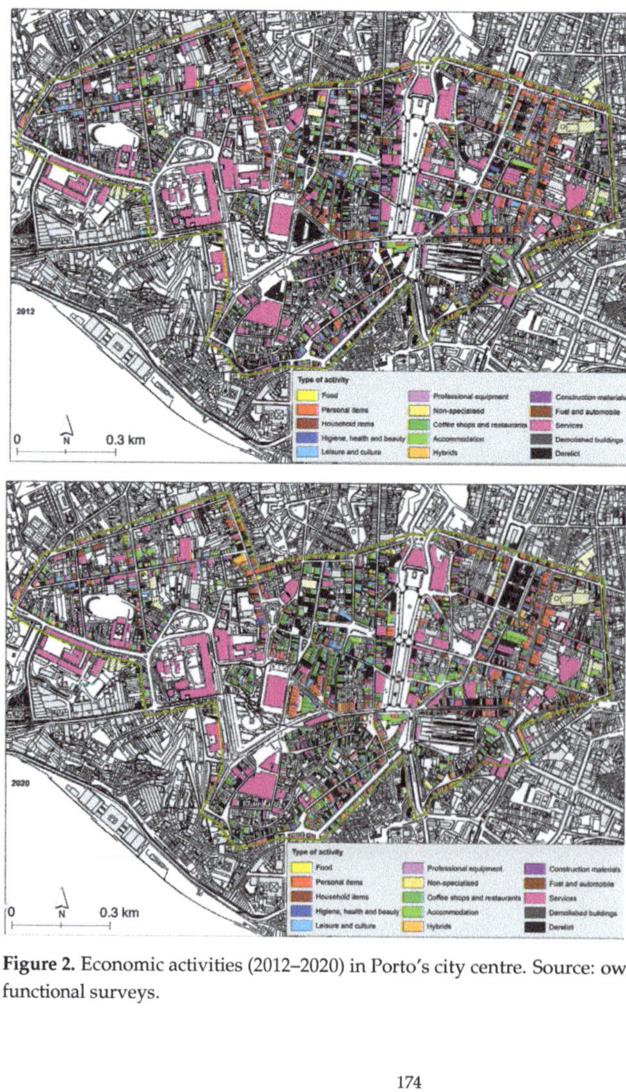

Figure 2. Economic activities (2012–2020) in Porto's city centre. Source: own elaboration, based on functional surveys.

In the interviews, when asked to indicate three options (out of 13) as priorities for urban policies (Table 1), it was found that economic development and housing collected 29 and 26 preferences, respectively, followed by "education, health and culture" and "accessibility and mobility" (both with 21).

Table 1. Porto downtown economic stakeholders' political priorities.

Political Priority	No. of Replies
Housing	29
Economic development	26
Accessibility and mobility	21
Education, health, and culture	21
Quality of life of the populations	18
Efficient and quality governance	13
Urban planning	9
Social support	8
Engagement of citizens	5
Demographic dynamics	4
Environmental sustainability	4
Justice	4
Territorial cohesion	0

This assessment was confirmed in the following answer, in which the respondent was asked to assess, from "much worse" to "much better", ten dimensions: safety, architecture, economy, housing, identity, sustainability, cleanliness, entertainment, neighbourhood and air quality (Figure 3). Here, housing stood out as the only domain that gathered the large majority of negative responses (39, with 23 of "much worse"). In all the others, except in relation to noise and safety, an improvement was noted, specifically in architecture (32 positive, 5 negative and 17 intermediate assessments), the economy (30 positive v 12 negative +12) and entertainment (27 positive v 12 negative +15).

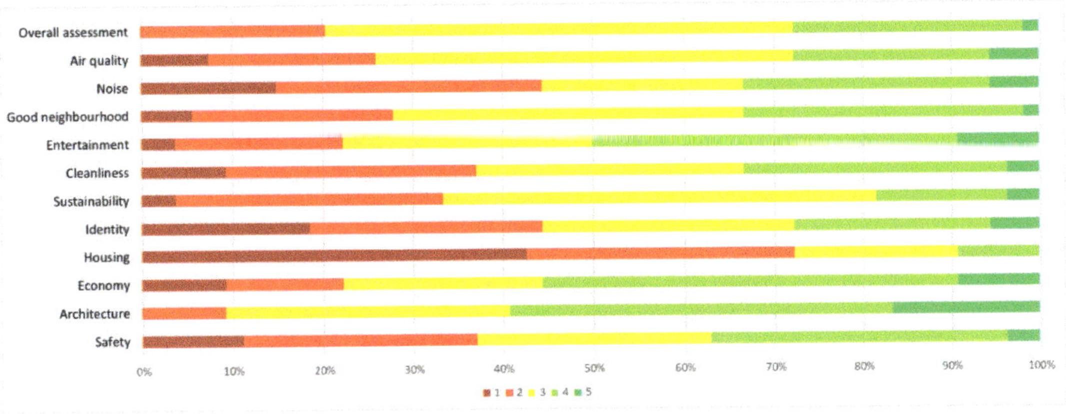

Figure 3. Economic stakeholders' assessment of the transformation at Porto's city centre. Source: own elaboration, based on interviews.

In the explanation for the transformations experienced in Porto's Baixa in the last decade (Figure 4), the role of low-cost flights (47 consider them important or very important), hotels (41), local accommodation (41) and real estate business (34) can be highlighted. Not so many references were made to urban rehabilitation (26) and the role of municipal policies (25). The lowest positive values were regarding housing policies (7) and social cohesion policies (12).

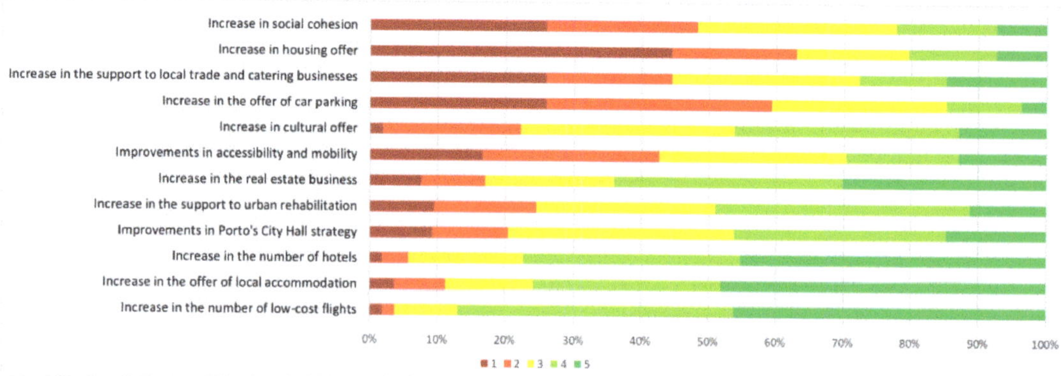

Figure 4. Economic stakeholders' assessment of the importance of the changes at Porto's city-centre. Source: Own elaboration, based on interviews.

During the interviews, the idea that there was a very important transformation and that tourism was at the centre of such change was universal. Tourism was referred to as "crucial", "essential", "vital", or "fundamental". But several of the respondents highlighting the positive effects on the economy and urban rehabilitation considered "expulsion" and "high rents." to be negative. Some critical voices were heard in a wider sense: "good for hotels and restaurants only", "embellished the city, but emptied it of residents", with several references to "de-characterisation" and "loss of identity".

When asked how Baixa should be, the most common references mentioned more residents, dynamism, safety, cleanliness, accessibility, justice, and green spaces.

On public policies, there was some embarrassment and contradictory ideas. Mentions included concerns with the homeless, the promotion of tourism and the struggle against gentrification, the need to balance tourism with the retention of residents and the importance of the existence of establishments capable of maintaining the identity of the city. "We need to have restaurants with Porto's traditional tripe stew", one of the interviewees said, because "the city cannot be like the 'malhão' (popular song), everyone "eating and drinking, walking in the street..." in the midst of nice facades (Figure 5).

Figure 5. (**Left** to **right** and **top** to **bottom**)—an old façade waiting for new buildings in the back (at Sá da Bandeira and Formosa streets); a tuc tuc and a lighted building with a shop that sells canned sardines on the ground floor; souvenir shops at Rua da Assunção; São Domingos in the evening, one of many places where restaurants, museums and other "attractions" bring tourists. (Own photos).

Finally, we sought to obtain indications regarding concrete proposals. Here, the most mentioned were the measures for more and cheaper housing (5), the improvement in collective transportation (5), the limitation of hotels and local accommodation (4), the creation of more cultural events and initiatives (4), the restriction of access to cars (4), more parking (3), further cleanliness (3), support for retail (3), video surveillance (3), free parking (2), end of evictions (2), improvement of public spaces (2), limitation of restaurants (1), social support for toxic dependents (1), increased parking for residents (1), more police in the street (1), gardens (1), spaces for senior citizens (1), day-care centres (1) and pedestrian streets (1).

5. Discussion

Tourism was regarded as the industry of peace. This was how it was promoted in the late 1940s in Europe, when it was perceived as a generator of tolerance and an agent in the restructuring of war-torn economies, prescribed to both developed and developing countries. However, the first signs of objection and prudence related to this reductive, almost magical, vision of tourism quickly emerged. After "Tourism: Blessing or Blight?" [46], the discussion has been kept alive to the present day regarding the two sides of the coin of tourism, while incorporating new concepts and clothing, which, in a way, killed its exoticism [47].

Enjoying tourism means moving away from everyday life. The massification of tourism, the new collaborative and deregulated types of the activity, and the behaviour, often gregarious, pose a threat to the sustainability of some of the most sought-after areas or even to an entire city. Even sustainable tourism is much talked, tourism normally has an important environmental impact [11,18] and its seasonality, vulnerability to crises, and dependence from travel causes a set of perverse effects.

Considering the discussions on the limits of tourism—how much is too much?—, the emergence of tourismphobia and tourism-led gentrification [48], a new model gained general acceptability, sustainable urban development, with social and spatial justice [49], sustainability [50,51] and economic innovation [52] as the main pillars. At the same time, the EU started to support cities in tackling urban decline through explicit initiatives and programmes, thus shaping a new area-based approach, characterised by strong coordination of action, horizontal partnerships, and the concentration of funding in specific vulnerable target areas.

In Porto, after decades of house rent freezing and the incapacity of public and private sectors to promote the rehabilitation of private buildings, as well as to maintain the attractiveness of the city centre, a liberal context (both in the central and local governments, after 2002) and especially the tourism explosion of the last decade promoted a radical change, as the statistics of land use evolution and interviews demonstrate. That also brought a significant alteration in the image and character of the city resulting from urbanistic and architectonic interventions, with an emerging new concern being associated with the excessive role of tourism in the future of the city—and especially in Baixa—and its relation with the principles of sustainability. In all the cases in which hotels or apartments for short rentals pretend to instal in vacant buildings, they rarely find any resistance. When there is previous use, there is a negotiated compromise or residents with older contracts and more than 65 can stay. But there are several cases of forced eviction denounced by grassroot movements, newspapers and posters on some buildings, with no success. Tourism is to reign in (neo)liberal cities. The resulting massification of tourism in Baixa is perceived in the interviews, where the passive role of public authorities (e.g., short-term rental is free, all year, everywhere, with no limits to the number of flats you rent, their prices, or the number of days you rent them, no matter whom) is also recognized. Tourism becomes not only central but almost the only "raison d'être" of the central city, its economy, animation and urbanism. As a consequence, the strong changes in progress in the centre of the city indicate a process of touristification, and no concern is expressed by the authorities about the negative effects of these transformations nor the potential of public policies to promote more sustainable tourism.

One of the main results is the intensification of the touristification of Porto's Baixa, thus corroborating the results of previous studies, which have documented, in many ways, the transformations taking place in the city centre [3,4,45,53–56].

What may be the role of Baixa in a more sustainable city with more sustainable tourism? We have seen that the last decades have been torn by successive crises and how a territorial specialisation in tourism can be a severe fragility, especially for a city centre [57,58].

However, despite the evidence of the strong transformations in progress and signs of fragility regarding the future of the city centre, it was possible to verify the existence of divergent opinions by the economic agents interviewed regarding the impacts of these transformations. This is indicated, from the outset, by the recognition of the key role of tourism in the transformations of Porto's Baixa in the last decade and by the multifaceted assessment of these transformations, for example, observable in the contrast between, on the one hand, the positive assessment in the economy, entertainment and rehabilitation of the buildings and, on the other hand, the negative consequence in housing. These results are in line with the already-recognised "double-edged nature" of tourism [25,31].

In fact, tourism has a multifaceted nature, and it is hard to individualise and manage the complex interactions and conflicting interests. Furthermore, in the specific case of Baixa, after a long crisis in the city centre, it is easier to understand that a valuation of the positive aspects of tourism prevails, which brings us to the idea that its acceptance or challenge will not depend only on scale and its proportions, but also on the social, economic, political, and institutional context [59]. In fact, in Porto, tourism has had a pivotal effect on the disruption of a long cycle of devaluation of the city centre, which is particularly evident in the continuous decline in population, urban de-qualification and the strong expression of derelict buildings. The memory of the economic crisis of the years 2007 and 2008, which was long and intense and impacted the independent shops in the city centre in a very particular manner, must be also important in the assessment. On the other hand, and similar to what occurred in many other cities, the increase in prices and the replacement of housing units by hotels or short-term accommodation, and their effect on the expulsion of residents and the increase in the difficulty of settlement, has led many to negatively assess changes and to identify this domain as a political priority in the city of Porto.

Finally, it is important to consider the interviews in relation to public policies. The interviewees' perception that tourism simultaneously stimulates positive and negative impacts on the city and on the quality of life of its inhabitants will certainly be one of the explanatory factors for the difficulties felt in envisioning the potential of public policies in the construction of a better city centre. On the other hand, the answers recognise the limited role of public policies in the management of tourism in the city, which is well evidenced in the residual weight attributed to policy measures (e.g., municipal policy, housing policy, and urban rehabilitation policy) in explaining the transformations experienced in the last decade. The results may perhaps be corroborating the acknowledged difficulty of devising public policies capable of responding to a complex system [25,31], which involves multiple agents from different fields and scales, and in which it is difficult to untangle the web between tourism, the city and the well-being of its residents.

But to address the challenges of overtourism in Porto while fostering sustainable urbanism, a multifaceted approach is possible, and seems essential. Firstly, zoning regulations that limit short-term rental, mitigating residence prices, evictions, and gentrification should be put in place. Establishing a balance between tourist accommodations and local housing is also crucial to achieve a long-standing diverse and sustainable community. Additionally, the necessity for new housing or rehabilitation to be mostly mixed-use, combining residential spaces with cultural and commercial areas, could help to reinforce the complexity of the city.

To preserve Porto's heritage and support local businesses, policies favouring historical shops and traditional retail in general as well as local artisans and regional products over generic tourist-oriented establishments and goods can help maintain the city's character.

In the same direction, work on buildings should transform the minimum and involve the least amount of production of construction materials as possible, as well as travel for goods and workers.

Strategic urban planning with a sustainable perspective should prioritize public spaces and pedestrian-friendly zones, and a reduction in the environmental impact of circulation. Investing in efficient public transportation systems, cycling infrastructure, and walkable areas not only eases congestion but also promotes eco-friendly travel.

Furthermore, fostering community engagement through participatory decision-making processes empowers residents, and their ideas will generally contribute to more sustainable urban development. The collaboration of local organizations, residents, and businesses in eco-tourism initiatives, such as special guided tours and community-led cultural events, are a good example of how local participation can enhance visitor experiences while minimizing environmental impact.

In short, a comprehensive strategy encompassing regulatory measures, thoughtful urban planning, community involvement, and responsible tourism promotion is crucial for ensuring Porto's sustainability as a tourism-dependent city, as well as intermunicipal coordination.

6. Conclusions and Comments

The research carried out in the city of Porto allows us to understand better the causes, characteristics and consequences of recent change in Baixa. We realised that the transformation of Porto's city centre, much oriented towards the tourist, has been promoting a process of residential and functional gentrification, as Carvalho, Chamusca, Fernandes and Pinto [54] noted. The multiplication of a new type of retail and service unit is associated with the increase in the price of the land (resulting from an increased tourist demand and urban rehabilitation processes, as well as the legal easing of the tenancy law), thus triggering the closure of many establishments and the displacement of others. The increase in tourist demand has also reinforced the real estate market attractiveness, leading owners (much of them new, and some being international funds) to significantly increase rents for housing, which, together with changes in the law, has led to evictions and a transformation of the social and economic pattern of the resident, with a significant increase in the non-permanent inhabitant. In fact, the National Institute of Statistics data show that the real estate selling price has grown 38.4% in the municipality of Porto and 44.7% in the historical centre parish between 2019 and 2022 (Portugal average was a 35.6% increase).

Tourism is identified as the main engine of transformation in the city centre. The vast majority of respondents highlight its effects on the economy and urban rehabilitation. However, they also notice that this urban revitalization came at a very high cost, particularly social, including the departure of many residents and small entrepreneurs, due to the widespread increase in land cost and the general cost of living.

Finally, the need for an integrated approach is evident from the interviews. The COVID-19 pandemic emphasised the city's enormous dependence on tourism and the unsustainability of this model of high specialization. As a consequence, Porto is a good reference for reflection on the post-COVID relationship between the city and tourism. The conclusion, supported by the interviews, allows us to argue for the advantage of a demand geared towards visits which are longer in time and calmer in speed. And, on the city side, for the need of an urban policy less eager to capture revenue and promote infinite inflows. These guidelines could be the foundation of a policy that on the one hand, prevents the mono-functionality of a "city to sleep, eat, drink and photograph" [60] (p. 53), and on the other hand, avoids the continuous increase in the number and distance of commutes, resulting from the residents' need to seek affordable residence, with the low salaries they earn in the establishments that tourism creates, and, finally, responds to the need for more sustainable tourism and a sustainable city, responding to the increase in avoidances of tourists from the most famous attractions [61].

More than a policy for tourism or a policy for the central city, the city of tourist consumption, in the face of the paradigm of sustainability, forces us to consider control over areas of overtourism and a multiscale and integrated perspective. A sound strategy will only be achieved if it is designed to articulate the centre with the entire urban area, and a strategy for tourism, in any territory and for the most part in a city, must be part of a more comprehensive strategy that considers interests other than just economic or short-term.

Author Contributions: Conceptualization, J.F., H.M., J.P., P.C. and R.L.; methodology, J.F., J.P., J.M. and P.C.; investigation, J.F., J.P., J.M. and P.C.; writing—original draft preparation, H.M.; writing—review and editing, J.F.; visualization, P.C.; supervision, J.F.; project administration, J.F. All authors have read and agreed to the published version of the manuscript.

Funding: This work received support from the Centre of Studies in Geography and Spatial Planning (CEGOT), funded by national funds through the Foundation for Science and Technology (FCT) under the reference UIDB/04084/2020, as well as from FCT financed projects UIDB/04084/2020, UIBD/00736/2020, and PTDC/GES-URB/30551/2017.

Data Availability Statement: Data is contained within the article.

Conflicts of Interest: The authors declare no conflicts of interest.

References

1. Glass, R.L. *London: Aspects of Change*; MacGibbon & Kee: London, UK, 1960.
2. Lees, L.; Phillips, M. *Handbook of Gentrification Studies*; Edward Elgar Publishing: Cheltenham, UK, 2018.
3. Chamusca, P.; Fernandes, J.R.; Carvalho, L.; Mendes, T. The role of Airbnb creating a "new"-old city centre: Facts, problems and controversies in Porto. *Boletín Asoc. Geógrafos Españoles* **2019**, *83*, 1–30. [CrossRef]
4. Fernandes, J.A.R.; Tenreiro, J.P.; Figueiredo, P.M.; Chamusca, P.; Pinto, J.R. *A Baixa do Porto: Arquitetura e Geografia Urbana (2010–2020)*; Book Cover: Porto, Portugal, 2021.
5. Scott, A.J.; Soja, E.W. *The City: Los Angeles and Urban Theory at the End of the Twentieth Century*; University of California Press: Los Angeles, CA, USA, 1996.
6. Lois González, R.C.; González Pérez, J.M.; Escudero Gómez, L.A. *Los Espacios Urbanos. El Estudio Geográfico de la Ciudad y la Urbanización*; Cátedra: Madrid, Spain, 2012.
7. Troitiño Vinuesa, M.Á. *Cascos Antiguos y Centros Históricos*; Ministerio de Obras Públicas y Transportes: Madrid, Spain, 1992.
8. Fernandes, J.A.R.; Sposito, M.E. *A Nova Vida do Velho Centro nas Cidades Portuguesas e Brasileiras*; CEGOT-Universidade do Porto: Porto, Portugal, 2013.
9. Chandler, A. What Should the 'Sharing Economy' Really Be Called? Available online: https://www.theatlantic.com/business/archive/2016/05/sharing-economy-airbnb-uber-yada/484505/ (accessed on 3 December 2023).
10. Frenken, K.; Schor, J. Putting the sharing economy into perspective. *Environ. Innov. Soc. Transit.* **2017**, *23*, 3–10. [CrossRef]
11. Blanco Romero, A.; Blázquez Salom, M.; de la Calle Vaquero, M.; Fernández Tabales, A.; García Hernández, M.; Lois González, R.C.; Mínguez García, M.C.; Navalón García, R.; Navarro Jurado, E.; y Troitiño Torralba, L. *Diccionario de Turismo*; Cádedra: Madrid, Spain, 2021.
12. Buhalis, D.; Amaranggana, A. Smart Tourism Destinations. In *Information and Communication Technologies in Tourism 2014*; Xiang, Z., Tussyadiah, I., Eds.; Springer International Publishing: Cham, Sitzerland, 2013; pp. 553–564.
13. Fernandes, J.A.R.; Carvalho, L.; Chamusca, P.; Gago, A.; Mendes, T. *Lisboa e a Airbnb*; Book Cover: Porto, Portugal, 2019.
14. Fernandes, J.A.R.; Carvalho, L.; Chamusca, P.; Mendes, T. *A Airbnb em Portugal*; Book Cover: Porto, Portugal, 2019.
15. González-Pérez, J.M. Una geografía de la gentrificación en el centro histórico de Palma (Mallorca, España). Turismo y elitización social. *Confins* **2020**. [CrossRef]
16. Yrigoy, I. Rent gap reloaded: Airbnb and the shift from residential to touristic rental housing in the Palma Old Quarter in Mallorca, Spain. *Urban Stud.* **2019**, *56*, 2709–2726. [CrossRef]
17. Wachsmuth, D.; Weisler, A. Airbnb and the rent gap: Gentrification through the sharing economy. *Environ. Plan. A Econ. Space* **2018**, *50*, 1147–1170. [CrossRef]
18. Milano, C.; Novelli, M.; Cheer, J.M. Overtourism and Tourismphobia: A Journey Through Four Decades of Tourism Development, Planning and Local Concerns. *Tour. Plan. Dev.* **2019**, *16*, 353–357. [CrossRef]
19. Pons, G.X.; Blanco-Romero, A.; Navalón-García, R.; Troitiño-Torralba, L.; y Blázquez-Salom, M. *Sostenibilidad Turística: Overtourism vs. Undertourism*; Societat d'Història Natural de les Balears: Palma, Spain, 2020.
20. León-Gómez, A.; Ruiz-Palomo, D.; Fernández-Gámez, M.A.; García-Revilla, M.R. Sustainable Tourism Development and Economic Growth: Bibliometric Review and Analysis. *Sustainability* **2021**, *13*, 2270. [CrossRef]
21. Cocola-Gant, A.; Gago, A. Airbnb, buy-to-let investment and tourism-driven displacement: A case study in Lisbon. *Environ. Plan. A Econ. Space* **2021**, *53*, 1671–1688. [CrossRef]
22. Saarinen, J. Critical Sustainability: Setting the Limits to Growth and Responsibility in Tourism. *Sustainability* **2014**, *6*, 1. [CrossRef]

23. Hall, C.M.; Gössling, S.; Scott, D. *The Routledge Handbook of Tourism and Sustainability*; Routledge: Abingdon, UK, 2015.
24. Aall, C.; Koens, K. The Discourse on Sustainable Urban Tourism: The Need for Discussing More Than Overtourism. *Sustainability* **2019**, *11*, 4228. [CrossRef]
25. Ashworth, G.; Page, S.J. Urban tourism research: Recent progress and current paradoxes. *Tour. Manag.* **2011**, 32, 1–15. [CrossRef]
26. Maxim, C. Sustainable tourism implementation in urban areas: A case study of London. *J. Sustain. Tour.* **2016**, *24*, 971–989. [CrossRef]
27. Shin, H.B. Planetary gentrification: What it is and why it matters. *Space Soc. Geogr. Thought* **2019**, 22, 127–137.
28. Bertocchi, D.; Camatti, N.; Giove, S.; van der Borg, J. Venice and Overtourism: Simulating Sustainable Development Scenarios through a Tourism Carrying Capacity Model. *Sustainability* **2020**, *12*, 512. [CrossRef]
29. Koens, K.; Postma, A.; Papp, B. Is Overtourism Overused? Understanding the Impact of Tourism in a City Context. *Sustainability* **2018**, *10*, 4384. [CrossRef]
30. Lerario, A.; Di Turi, S. Sustainable Urban Tourism: Reflections on the Need for Building-Related Indicators. *Sustainability* **2018**, *10*, 1981. [CrossRef]
31. Fistola, R.; La Rocca, R. Driving functions for urban sustainability: The double-edged nature of urban tourism. *Int. J. Sustain. Dev. Plan.* **2017**, *12*, 425–434. [CrossRef]
32. Roggema, R. The future of sustainable urbanism: A redefinition. *City Territ. Archit.* **2016**, 3, 22. [CrossRef]
33. Moreno, C. *Droit de Cité: De la" Ville-Monde" à la" Ville du Quart d'heure"*; Editions de l'Observatoire: Paris, France, 2020.
34. Allam, Z.; Bibri, S.E.; Chabaud, D.; Moreno, C. The '15-Minute City' concept can shape a net-zero urban future. *Humanit. Soc. Sci. Commun.* **2022**, *9*, 126. [CrossRef]
35. LeGates, R.T.; Stout, F. *The City Reader*; Routledge: London, UK, 1996.
36. Haase, D. Integrating Ecosystem Services, Green Infrastructure and Nature-Based Solutions—New Perspectives in Sustainable Urban Land Management. In *Sustainable Land Management in a European Context: A Co-Design Approach*; Weith, T., Barkmann, T., Gaasch, N., Rogga, S., Strauß, C., Zscheischler, J., Eds.; Springer International Publishing: Cham, Switzerland, 2021; pp. 305–318.
37. Bibri, S.E.; Krogstie, J.; Kärrholm, M. Compact city planning and development: Emerging practices and strategies for achieving the goals of sustainability. *Dev. Built Environ.* **2020**, *4*, 100021. [CrossRef]
38. Nieuwenhuijsen, M.J. New urban models for more sustainable, liveable and healthier cities post covid19; reducing air pollution, noise and heat island effects and increasing green space and physical activity. *Environ. Int.* **2021**, *157*, 106850. [CrossRef] [PubMed]
39. Hall, P. *Cities of Tomorrow: An Intellectual History of Urban Planning and Design in the Twentieth Century*; Blackwell: Oxford, UK, 1988.
40. Hall, P. Forces Shaping Urban Europe. *Urban Stud.* **1993**, *30*, 883–898. [CrossRef]
41. Scott, A.J. Globalization and the Rise of City-regions. *Eur. Plan. Stud.* **2001**, *9*, 813–826. [CrossRef]
42. Le Galès, P. Urban Policy in Europe. In *International Encyclopedia of the Social & Behavioral Sciences*, 2nd ed.; Wright, J.D., Ed.; Elsevier: Oxford, UK, 2015; pp. 900–907.
43. Faludi, A. A turning point in the development of European spatial planning? The 'Territorial Agenda of the European Union' and the 'First Action Programme'. *Prog. Plan.* **2009**, *71*, 1–42. [CrossRef]
44. Parkinson, M. Urban policy in Europe: Where have we been and where are we going. In *European Metropolitan Governance: Cities in Europe-Europe in Cities*; Antalowsky, E., Dangschat, J., Parkinson, M., Eds.; Proyecto NODE (Ministerio Austriaco de Educacion, Ciencia y Cultura): Vienna, Austria, 2005; pp. 17–67.
45. Fernandes, J.A.R.; Carvalho, L.; Chamusca, P.; Mendes, T. *O Porto e a Airbnb*; Book Cover: Porto, Portugal, 2018.
46. Young, G. *Tourism: Blessing or Blight?* Penguin: Harmondsworth, UK, 1973.
47. Hughes, G. Turismo, sustentabilidade e teoria social. In *Compêndio de Turismo*; Lew, A., Hall, C., Williams, A., Eds.; Instituto Piaget: Lisboa, Portugal, 2007; pp. 557–569.
48. Cocola-Gant, A. Gentrificación turística. In *Turísmo Global. Perspetivas Críticas Sobre el Turismo*; Cañada, E., Murray, I., Eds.; Icaria: Barcelona, Spain, 2019.
49. Soja, E.W. *Seeking Spatial Justice*; University of Minnesota Press: Minneapolis, MN, USA, 2010.
50. Camagni, R. Regional Competitiveness: Towards a Concept of Territorial Capital. In *Modelling Regional Scenarios for the Enlarged Europe: European Competiveness and Global Strategies*; Capello, R., Camagni, R., Chizzolini, B., Fratesi, U., Eds.; Springer: Berlin/Heidelberg, Germany, 2008; pp. 33–47.
51. Pires, S.M. Urban Sustainability. In *The Wiley—Blackwell Encyclopedia of Social Theory*; Turner, B.S., Ed.; John Wiley & Sons: Hoboken, NJ, USA, 2017.
52. Ranci, C. Competitiveness and Social Cohesion in Western European Cities. *Urban Stud.* **2011**, *48*, 2789–2804. [CrossRef]
53. Alves, S.; Branco, R. Models of urban rehabilitation under neoliberalism and austerity: The case of Porto. In *Spaces of Dialog for Places of Dignity—Fostering the European Dimension of Planning: Book of Proceedings*; Freitas, C., Fonseca, I., Eds.; Universidade de Lisboa: Lisbon, Portugal, 2017; pp. 1839–1851.
54. Carvalho, L.; Chamusca, P.; Fernandes, J.; Pinto, J. Gentrification in Porto: Floating city users and internationally-driven urban change. *Urban Geogr.* **2019**, *40*, 565–572. [CrossRef]
55. Fernandes, J.A.R. *Geografia do Porto*; Book Cover: Porto, Portugal, 2020.
56. Ferreira, C.M.F. Metamorfoses da Baixa do Porto. Dinâmicas Económico-Sociais e Estratégias Públicas e Privadas em Contexto Urbano na Última Década. Ph.D. Thesis, University of Porto, Porto, Portugal, 2018.

57. Fletcher, R.; Murray Mas, I.; Blanco-Romero, A.; Blázquez-Salom, M. Tourism and degrowth: An emerging agenda for research and praxis. *J. Sustain. Tour.* **2019**, *27*, 1745–1763. [CrossRef]
58. Simancas Cruz, M.; Hernández Martín, R.; Padrón Fumero, N. *Turismo pos-COVID-19: Reflexiones, Retos y Oportunidades*; Universidad de La Laguna: La Laguna, Spain, 2020.
59. Novy, J.; Colomb, C. Urban Tourism as a Source of Contention and Social Mobilisations: A Critical Review. *Tour. Plan. Dev.* **2019**, *16*, 358–375. [CrossRef]
60. Fernandes, J.A.R.; Chamusca, P. As alterações do contexto e o triunfo do turismo. In *A Baixa do Porto: Arquitetura e Geografia Urbana (2010–2020)*; Fernandes, J.A.R., Tenreiro, J.P., Figueiredo, P.M., Chamusca, P., Pinto, J.R., Eds.; Book Cover: Porto, Portugal, 2021; pp. 44–53.
61. Larsen, J. Ordinary tourism and extraordinary everyday life: Rethinking tourism and cities. In *Tourism and Everyday Life in the Contemporary City*; Frisch, T., Sommer, C., Stoltenberg, L., Stors, N., Eds.; Routledge: London, UK, 2019; pp. 24–41.